Jaroslav Flegr, Professor in Ecology, graduate of the Faculty of Science of Charles University in Prague, studied cell biology and physiology and has devoted most of his scientific career to molecular phylogenetics and evolutionary parasitology. He is the author of a large textbook Evolutionary Biology (Academia, Praha 2005) and more than 60 research papers, mostly on evolutionary biology and parasitology. The Czech version of his book Frozen evolution was nominated for a title "Book of the Year" in 2006 and was awarded a prestigious literature award Magnesia Litera in 2007.

Jaroslav Flegr

Frozen Evolution

Or, that's not the way it is, Mr. Darwin
Farewell to selfish gene

Prague 2008
Charles University in Prague, Faculty of Science

This book is dedicated to the memory of the two most influential evolutionary biologists of the end of the 20th century, S.J. Gould and J. Maynard Smith, the synthesis of whose theories (punctuated equilibrium and evolutionarily stable strategy) led to the theory of frozen plasticity.

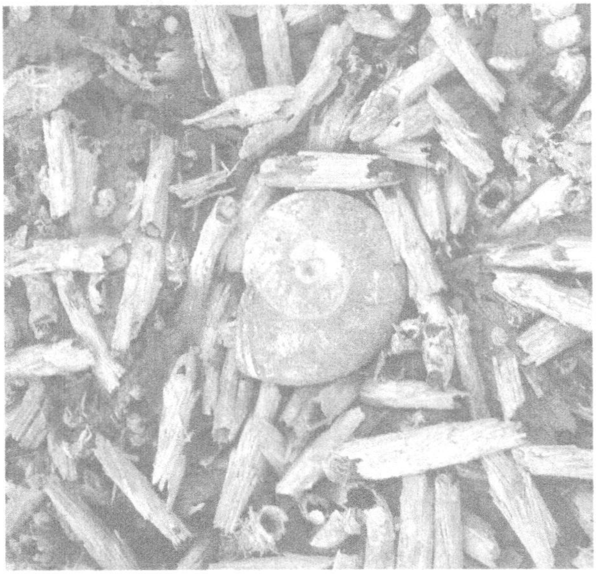

Flegr Jaroslav
Frozen Evolution. Or, that's not the way it is, Mr. Darwin – Farewell to selfish gene

Printed and bound in United States of America.

ISBN: 978-80-86561-73-8

Contents

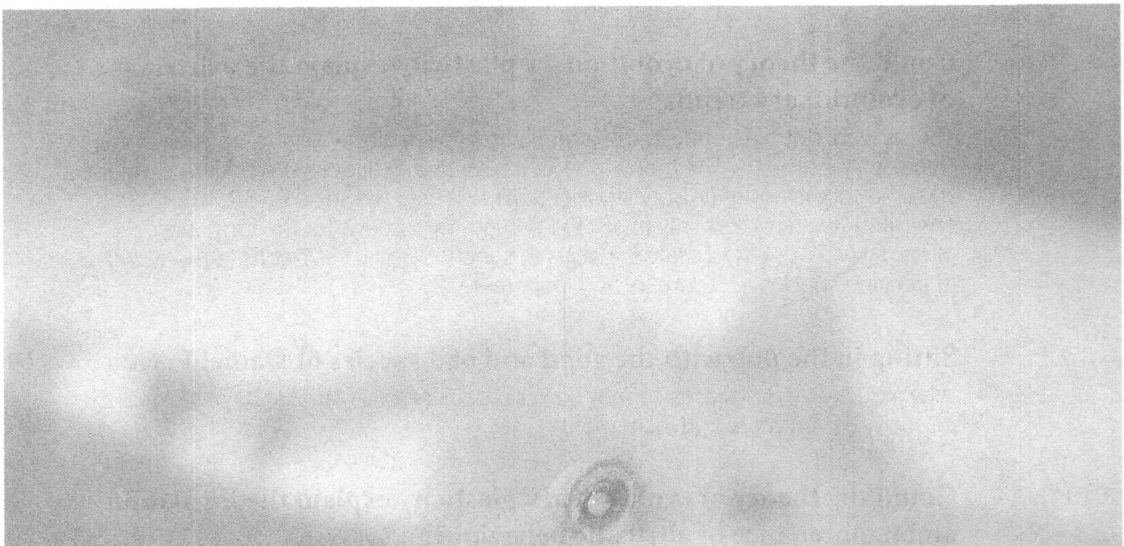

CHAPTER 1 Everything is not the way it sounds

It is already almost 150 years since the first edition of the book "On the Origin of the Species by means of Natural Selection" was published.[1] In this work, the most famous biologist of all time, Charles Darwin, was the first to write and give the reasons why animal and plant species originated and continue to change in nature by natural development from a common ancestor. He explained that the diversification of the species is caused by natural selection, preferential survival and multiplication of the fittest individuals. Natural selection thus simultaneously satisfactorily explains the useful adaptation of organisms to the conditions in their environment. Darwin's ideas encountered strong resistance in the religiously oriented society of that time; however, they were almost immediately adopted in scientific circles. Over time, his theory of evolution has been confirmed innumerable times over and, to the present day, remains the basis for all scientific evolutionary theories.

I assume that the reader is not in any way offended by any of the statements in the previous paragraph. Nonetheless, with the exception of the rather uninteresting statement that it has been almost 150 years since the publication of the first edition of Darwin's book, everything else is basically untrue. Charles Darwin (1809–1882) was certainly not the first biologist to submit to the professional public a comprehensive theory of development of the species by gradual evolution from a common ancestor, as Jean-Baptiste Lamarck (1744–1829) had done this 50 years previously in his key work "Zoological Philosophy".[2] The driving force for the diversification of the species is quite possibly not natural selection, but an entirely different evolutionary mechanism, which will be discussed in Chapter 4. Natural selection, as described by Darwin, permitted explanation of the formation of adaptive traits in bacteria, but is inadequate for explanation of these traits in the vast majority of "higher" organisms.[3] Darwin's ideas were accepted with surprising favour by the general public. In contrast, over time, a number of biologists have put forth and continue to put forth relatively serious objections against them. Truly, Darwin's theory was subsequently confirmed many times over. However, simultaneously, a number of facts emerged over time that threw its validity into doubt in many respects, or at least restricted its validity. In the light of modern knowledge, Darwin's model can no longer be considered as forming the basis of scientific theory related to the formation and development of life. This book will be concerned with this subject and with a theory that could replace Darwin's theory.

I want to make it perfectly clear right from the beginning: not to disparage Darwin. Similar to him and the vast majority of my colleagues, I too am convinced that organisms were formed over the extremely long duration of the Earth, one from another by the natural process of **biological evolution**.

Compared to Darwin, I have a far greater factual basis for my convictions, knowledge

Box 1.1 Evolution

Evolution is generally understood to mean gradual development of any system with a "memory", i.e. any system that responds to external stimuli in dependence on the stimuli that it encountered in the past. This means that it is equally possible to speak of the evolution of languages, automobile chasses or ladies' hairstyles as about the evolution of conifers. Evolution can be direct, reverse or cyclic. **Biological evolution** is one of the many types of evolution. It is interesting primarily in that organisms are formed spontaneously during this process, i.e. systems that are usefully adapted to the use of various resources in the environment, including such marvellous creatures as fruit flies, coconut palms, sturgeons and the readers of this book.

accumulated by biologists over the 150 years that have expired since the publication of "On the Origin of the Species by means of Natural Selection". However, in contrast to most proponents of traditional thinking in evolution biology, I am of the opinion that the manner of formation and development of species through the action of natural selection is different than that described by Darwin and that modern textbooks attempt to show us. I am further convinced that the differences between the Darwinian and new model of evolution have a fundamental impact on our understanding of the progress of a number of natural processes. Many of these processes, which are difficult to understand in the intellectual framework of the older theory of evolution and that actually occur, although they should not occur in a Darwinisitic world, can be relatively easily explained in the framework of the new model.

What is the actual basis for my heresy? **While Darwin's original theory assumed that the species that are encountered in nature are evolutionarily plastic and more or less willing to respond to the selection pressure of the environment – i.e. usefully adapt to its changes, the new theory[4] assumes, to the contrary, that the vast majority of species does nothing of the sort and, in fact, cannot do so.** These are species that I will call evolutionarily frozen in this book. These species respond to changes in their environment like rubber – initially they give in to the environmental pressure and change somewhat, however, the more their traits differ from the original state, the greater resistance they exert against the pressure until, at a certain point, they cease to react to even the greatest pressure. While, in a Darwinian world, all the species gladly develop and continuously change in response to ever newer demands from a changing environment, in a world with frozen plasticity, species remain more or less unaltered and mostly only sadly wait until the changes in their environment accumulate to such a degree that they will have no other alternative than to simply pass into extinction. Why this is true and where the new species come from, how it is possible that species are usefully adapted to their environment and how evolution can occur at all in such an evolutionarily frozen world – these are aspects that I do not intend to address right here. However, if you don't put this book down prematurely, answers will be provided to these questions.

I hope that I have managed to awaken the interest of readers in the previous paragraphs and that I can begin in the next chapter with an introductory presentation of the generally known Darwin's model of biological evolution. In the following chapters, we will gradually discuss the most important new aspects that Neodarwinism brought to Darwin's theory in the 20th century. Chapter 8 will be concerned with Dawkinistic evolution, i.e. the selfish gene theory. This is a model of evolution that should resolve the difficulties of the Neodarwinist theory with explanation of evolution in sexually reproducing organisms. This model[5] assumes that what seems to be Darwinian evolution is, in actual fact, a sort of puppet show held by the individual genes in the framework of their race to be the fastest to multiply – to produce copies of themselves. In the 9th and 10th chapter, we will show that the selfish gene theory also does not resolve a fundamental problem of the evolution of sexually reproducing organisms. If you read the book in the same way as some people I know, i.e. you begin to read somewhere in the middle of the book, then you skip over three chapters, you read all the adjectives in Chapter 16 and all the verbs in Chapter 17 and, if even then you still find the book interesting, you return to the chapter with the nicest pictures, I would recommend that you begin in Chapter 8 and then perhaps you could also run through Chapters 9 and 10. (I know that, in that case, you would not be reading this part or that you would read it at the very end; however, perhaps you accidentally opened the book here – so why not try.) However, you should definitely not skip over Chapter 11 which is of key importance from the standpoint of what I want to say. It describes a model of evolution that I think best corresponds to modern knowledge of evolutionary biology and palaeontology. The following seven chapters of the book gradually present fundamental facts supporting the validity of the new model and some of its interesting consequences. The last chapter contains a discussion of why Darwin's model of evolution persists in textbooks and what the chances are of a change and acceptance of the new model, which has been knocking on the door without visible results from as far back as the 1960s at the very least.

If the non-biologist reader occasionally loses his way in the text or, for example, doesn't understand a technical term (and the reference in the index is of no help), it doesn't matter too much. In this case, I would recommend that he simply keeps reading and it will probably all become clear in a short while. And one more technical comment for those who prefer to jump back and forth in a book. Each chapter ends with a paragraph that briefly and clearly summarizes its main message and indicates what will be the subject of the next chapter. It is clear that an addicted book hopper will hardly let himself be deprived of the experience of discovering his own way into the book by leafing through it (and, what is more, from the front to the back) and stopping at the conclusion of each of these concluding paragraphs. However, this approach could be a bearable concession for less addicted book hoppers.

The readers of popular educational books (including myself) are ridiculously pampered in the present day and age. Thus, it has become habitual to arrange the text so that more difficult sections alternate with very easy sections. It is optimal if the scientific passages are interspersed with recounting about scientific expeditions to exotic countries or description of the background of the individual discoveries, which should, wherever possible, contain amusing anecdotes about their participants. Unfortunately, I cannot include anything of this kind. I have, in fact, spent quite a bit of time on expeditions to exotic countries; however, I did not encounter any evolution there and it goes against the grain to present this fact as proof of the theory of frozen plasticity. As I tend to be an introvert and, in addition, spent half of my professional life on the wrong side of the iron curtain, I cannot offer my readers any impressions from my personal meetings with the important contemporary evolutionary biologists. I finally arrived at an approach that followed from my years of experience in lecturing for the course "Practical methodology of science" for students of the master's program in biology of the Faculty of Science of Charles University. I found from the response of students that some less known and less apparent aspects of scientific work seem quite interesting and even rather entertaining to listeners. Thus, why not share this with the general public, especially when this can help in placing the subject of interest in the broader framework of contemporary science.[6] However, you needn't be afraid, you can, but need not, accept the invitation to the exotic land of hypotheses, grants and impact factors. If you have no desire to look under the lid of contemporary science or if my point of view offends you, if will be sufficient to ignore the relevant grey boxes. Even though, here and there, I will slip in snippets of my experience but this will, I hope, be at an acceptable level.

Perhaps I should just mention my point of view briefly. I studied cell biology and have spent most of my scientific career studying molecular phylogenetics and evolutionary parasitology. Similar to my colleagues, I receive funds for my scientific projects from scientific grant agencies, regularly publish articles in international journals, give lectures at the university and supervise my master's and doctoral students. I am certainly no scientific dissident who, ignored by his surroundings, works somewhere away from the main stream of science, or a scientific celebrity who always swims in the centre of the main stream and is thus regularly invited as a plenary lecturer at scientific congresses. I am one of the many scientists who moves along somewhere between the extremes and, simultaneously, I am constantly aware of the great luck I have in a profession entailing activities that I enjoy above all – to discover that which has not yet been discovered. I will ever be grateful to the Gods (or blind chance) and to the tax payers for this.

The results of my rather unextensive statistical survey (I asked Frank and Charles) indicate that half of the readers leave out the preface and introduction in books. As most readers will have realized, I took note of this fact and took the necessary remedial measures. I craftily called the preface that you have just read "Chapter 1". It could well be that I have a number of these tricks up my sleeve. Amongst other things, I began to write the book "Frozen Evolution" to work off steam after seven years of writing horribly fat textbooks on evolutionary biology.[7] (I did this quite voluntarily and even enjoyed it from time to time, but there was truly a lot of it). Dearest readers, please be prepared for the possibility that I will occasionally act in a way that you may find too personal (for which I not entirely sincerely apologize). In fact, I can't completely eliminate the possibility that I could occasionally make fun of you. I don't mean by this that I would consciously slip in untrue information or intentionally disguise facts that I find unsuitable for my purposes. Selective memory is, of course, a bastard, so it will be better if you expect beforehand that I will tend to mention results that agree with my favourite theory rather than those that disagree with it. However, in this respect, my text will not differ substantially from the texts of other authors. Impartial, completely objective books probably don't exist and, if they did, they wouldn't be readable. Similarly, I admit beforehand that, in the role of the author of a popular educational book, I will have to act in a rather undisciplined manner – here and there, I will let the subject wander a bit more than is usual. For example, in the first half of the next chapter.

Footnotes

1. Of the great many editions of this famous book, I can recommend: *On the Origin of Species by Means of Natural Selection* or the *Preservation of Favoured Races in the Struggle for Life.* Murray (London) 1860.
2. Lamarck's book of 1809 was published in English, e.g. in 1984: *Zoological Philosophy. An Exposition with Regard to the Natural History of Animals.* The University of Chicago Press (Chicago and London).
3. "Higher organisms" is frequently used; however, from the standpoint of the theory of evolution, this is completely the wrong term. All the known organisms on the Earth developed from a common ancestor and thus have the same age, and they have all demonstrated their quality by surviving to the present day. Generally, this term is used to refer to organisms that are most closely related to us – human beings. This would mean that a chimpanzee is a higher organism than a dog, that a dog is a higher organism than a parrot, that a parrot is a higher organism than a trout, that a sea squirt is a higher organism than a martagon lily and that an amoeba is a higher organism than a bacterium. No one, of course, doubts that human beings are the highest organism and, at the present time, we needn't take into account the opinions of dolphins and octopuses. But some time in the future, who knows...?
4. I first published a paper on the theory of frozen plasticity in Flegr, J., On the "Origin" of natural selection by means of speciation. *Rivista di Biologia–Biology Forum* 91: 291–304, 1998.
5. The selfish gene theory was first published by Richard Dawkins in the popular book *The Selfish Gene*, Oxford University Press (Oxford), 1976. He published his concepts of evolution for professionals (but still in a very comprehensible form) in the book *Extended Phenotype, The Gene As the Unit of Selection.* W.H. Freemen and Comp. (Oxford), 1982.
6. Out of consideration for future readers, I first tested the effects of this manuscript on people in my vicinity, both scientists and representatives of other professions. It followed from the subsequent conversations that many of them considered it important to learn what is a "scientific fact" and what is scientific theory, and what is only a "supposition", who first discovered what and who what else, what is generally known and what is something I thought up. (In general, they exhibited a tendency to take seriously only what is generally known – which, I admit, somewhat annoyed me.) I think that, without the grey boxes in the text, it would be substantially more difficult to explain to them that a considerable part of their questions don't make sense, that, in short, things work differently in science.
7. If anyone were interested in submerging themselves in a fat book about evolutionary biology (in Czech), then they can do this: *Evoluční biologie*, Academia (Prague), 2005. Considering that I wrote it... (Some material in English is also available at http://natur.cuni.cz/flegr/book_evbiol.php).

Most of the footnotes were placed at the end of the chapters for technical reasons. However, I made an exception in this case. I would like to take this opportunity to thank all my colleagues, students, acquaintances and acquaintances of my acquaintances on whom I tested the manuscript over the past years. As there were about 30 of them and I don't know many of their names, I am, unfortunately, forced to thank them *en masse*. Thanks go to them for their many valuable comments and suggestions related both to the substantive and also to the formal aspects of the text. In addition, I would also like to thank, in advance, future readers of the book, who can send their comments to j@flegr.eu. Errata and other new material related to the text will be published at the web site: frozenevolution.com.

CHAPTER 2 The emergence of Darwinism or what Darwin did and did not discover and how

Anyone who has carefully read the book "On the Origin of Species by means of Natural Selection" soon discovers that Darwin was not only a brilliant biologist, but also a very capable author of scientific texts. The book is written very competently in relation to future readers and especially potential critics. When I explain to students how to write a scientific paper, the first thing I emphasize is the important principle that the "Introduction" chapter should be written last.

Box 2.1 Scientific article

At the present time, scientific results are usually published in the form of a brief article in one of the many thousands of scientific journals. A scientific article usually consists of a brief *Abstract* summarizing the most important results, of an *Introduction* chapter, which is intended to describe the purpose of the study and place it in the broader context of the field, of a *Material and Methods* chapter, containing the description of used methods, of a *Results* chapter, containing the uncommented results of the study (we measured this and that, the difference was/was not statistically significant), a *Discussion* chapter, stating what we think our results mean, how they agree or do not agree with knowledge to date and what follows from them. The article is usually ended with acknowledgement of people who contributed to completion of the study (but not enough to be included amongst the authors of the study) and of grant agencies that financed our research work, see Box 7.8 on p. 105, and also a list of references cited in the article, see Box 2.6 on p. 18. Overall, an article (in the fields of biology) usually has 2000–6000 words and 3–6 graphs and tables, i.e. takes 4–12 pages in the journal.

This chapter is intended to show the reader what the purpose of this study was and why it had to be carried out just now and in just this way. Science has a rather unpleasant characteristic from the standpoint of writing introductions: it does not always answer the question that was posed when the scientific work was commenced. It is frequently found that the greatest benefit from a project is a discovery that we made quite unexpectedly when studying entirely different phenomena. It will be most illustrative to demonstrate this statement on an example from the work in our laboratory.

How we accidentally discovered the function of the Rh factor

When, in our laboratory, we studied the effect of infection by the parasitic protozoa *Toxoplasma* on the abilities of infected persons, we discovered the not surprising fact that individuals with protozoa cysts in the nerve and muscle tissue react more slowly to simple stimuli.[1]

This is not surprising, because toxoplasma affects the abilities of infected mice in the same

Box 2.2 Toxoplasma

In developed countries, *Toxoplasma* may well be the most widespread protozoan human parasite. Its definitive host (i.e. the host in which it reproduces sexually) consists of a cat (any member of the family Felidae). The infected cat excretes resistant cysts (oocysts) into its environment in its excrement; these can enter an intermediate host with food or water. Any bird or mammal, including human beings, can be an intermediate host. *Toxoplasma* reproduce only asexually in the bodies of intermediate hosts and form the latent stage, tissue cysts, in the muscle and nerve tissue. If a cat catches an infected intermediate host and eats the tissue cysts, it becomes infected and the reproductive cycle of the parasite is closed. Human beings are most frequently infected by eating raw or improperly cooked meat or insufficiently washed vegetables contaminated with soil containing oocysts. In healthy humans, the infection has symptoms similar to those of common viral or bacterial diseases and rapidly disappears. However, viable cysts remain in the nerve and muscle tissue throughout life. If a pregnant woman becomes infected, the infection can (but need not) pass through the placenta to the foetus, which can be seriously damaged. Consequently, pregnant women should not eat incompletely cooked meat, soil-contaminated vegetable and should not clean cat litter boxes.

way.[2] However, another, quite unexpected discovery was far more important. We found that the infection has a different effect on the psychomotor performance (reaction times) of Rh negative and Rh positive persons. Amongst uninfected persons, the reaction times of Rh negative individuals were substantially better than those of Rh positive persons. However, the situation was the opposite in infected persons. As a consequence of the infection, the Rh negative persons had substantially reduced abilities, i.e. they exhibited substantially longer reaction times, while Rh positive persons had only slightly reduced or unaltered abilities. This could explain the origin of the Rh protein. In most countries, 20–80% of the population is infected with toxoplasma throughout their lives. In the past, the risk of infection was even greater, especially in areas with high occurrence of cat species. Africa is such an area and over 90% of the indigenous population is infected at the present time. If most of the people in the population were infected with this parasite, Rh positivity – increasing resistance to the unfavourable effects of the infection – could constitute a substantial advantage and the relevant gene could spread in the population. Thus, it need not be accidental that there are practically no Rh negative people amongst the indigenous inhabitants of Africa.

Box 2.3 Rh negative and Rh positive persons

People can be divided into two groups, differing in the presence of certain forms of a protein on the surface of the red blood cells. Rh positive persons (about 80% of the European population) have the relevant molecule on their red blood cells, while this molecule is missing in Rh negative persons (in actual fact, it is usually there, but is altered – however, that is not important here). If the blood of an Rh positive person is transferred to the body of an Rh negative person, the appropriate antibody molecules are formed and destroy the blood cells derived from the Rh positive person. Transfer of blood from an Rh positive person can occur during organ transplants or naturally in Rh negative women who expect an Rh positive child (with an Rh positive father). In the past, the presence of these antibodies seriously affected the lives and health of subsequent children of the same woman.

Box 2.4 Gene

This is one of the basic concepts of modern biology designating the predisposition for certain traits. However, even professionals in various fields cannot agree on a specific meaning for the word gene. Molecular geneticists have a clear concept in this respect as they define a gene as a continuous segment of a DNA molecule. Evolutionary biologists know this is ridiculous and that a gene cannot be defined in this way[3], but they are in a negligible minority at the present time and, if they were to fight for their version of the truth, they would definitely suffer defeat. Consequently, they prefer to grit their teeth in silence and act as if everything were fine (and usually alternately use the concept of a gene in the original and in the molecular biological meaning). In this entire book, the concept of gene could be replaced by the word predisposition. The reason why I don't satisfy non-biologist readers and why I don't replace it by the word predisposition lies primarily in the fact that I will also have to use other technical terms that are derived from the word gene. These include genotype, genome and gene pool. Predispositiontype, predispositionome, predisposition pool – no I guess that wouldn't work.

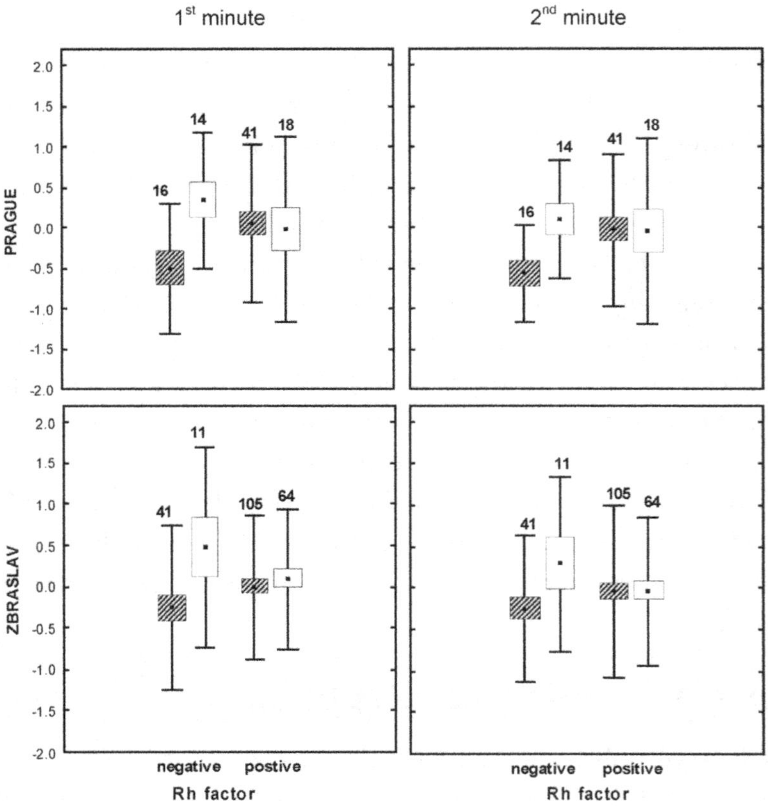

Fig. 2.1 Differences in the effect of infection by the *Toxoplasma* parasite on the reaction times of Rh negative and Rh positive men. Uninfected Rh negative men react fastest to simple instigations. However, following infection (blank boxes), their reaction times are substantially prolonged. Rh positive men react more slowly to simple instigations, but their reaction times practically do not decrease following infection. The reactions times are expressed in the graph in Z-scores (deviations from the average value) rather than in milliseconds, the spreads indicate mean standard deviation, boxes mean +/− standard error, figures indicate the numbers of *Toxoplasma*-free and *Toxoplasma*-infected men.

In contrast, in Europe, where large wild cats were always very scarce compared to Africa and where house cats were kept throughout the middle ages and for much of modern times only by witches, who managed to escape burning at the stake, toxoplasmosis must have been very rare. It is not surprising that a variant of the gene responsible for Rh negativity, i.e. the allele responsible for greater performance of persons not infected by toxoplasmosis, spread here.[4]

Let us return to the order of writing the individual chapters of a scientific article. A scientific article must completely truthfully inform the reader of the results obtained. However, it would be very unreasonable to write in the introduction that we were not at all interested in the Rh factor as the beginning, but that the original intention of our study was to test the **manipulation hypothesis**, specifically to determine whether the parasitic protozoa *Toxoplasma gondii* prolongs the reaction time of infected individuals (e.g. the speed with which they are capable of pressing a button after a little white square appears in the centre of the monitor of a computer), increasing the probability of its transmission from an intermediate host (under normal conditions, mice) to the digestive system of its definitive host – feline predators.

Box 2.5 Manipulation hypothesis

According to the manipulation hypothesis, a number of parasites purposefully and specifically alter the behaviour of their hosts and thus increase the probability of their transmission to an uninfected host. For example, it is assumed that toxoplasmosis can reduce fear of cats in infected rodents, or reduce the speed with which they can react to simple stimuli.[5] Parasites transmitted by sexual intercourse could increase the sexual activity of their hosts or the attractiveness of infected males for females. In some cases, parasites affect the behaviour of their hosts directly, e.g. by targeted interventions into the nervous system (rabies), in some cases indirectly; for example, the bacteria causing the plague damage the oral system of fleas so that they can bite, but cannot suck blood, an infected flea is therefore constantly hungry and bites and thus infects more hosts.

Because of the lack of space in our animal facilities and the shamelessly high prices of laboratory animals, we decided to use human volunteers instead of laboratory mice. We monitored the Rh factor only because we used volunteer blood donors for whom the relevant data was available.

No, no, no, of course not! We really couldn't admit that. Scientific discoveries should not be made by a combination of happy coincidences, but rather through a previously carefully prepared and targeted project (see Box 7.8 on p. 105). If the article were to have any chance at all of success under the review procedures of a scientific journal (see Box 2.7 on p. 19), it was necessary to write something completely different in the introduction. For example, that the presence of Rh negative and Rh positive persons in the human population is an evolutionary mystery and that the different sensitivity of carriers of the two variants of the relevant gene to the effect of some infection could be responsible for this. The candidate for the role of the origin of the infection is *Toxoplasma gondii*, which infected a large percentage of the population in the past. Then there would have to be a few references to the frequency of the occurrence of toxoplasmosis in the human population

Box 2.6 References

It is not possible to simply assert something in a scientific text; all our statements must have a basis. Either a reason must be given for our statement or we must demonstrate that someone made (and thus somehow justified) this statement before us. References are used for this second purpose – the name or names of the authors of the relevant source and the year of the publication are written directly in the text and a list of references is placed at the end of the text, giving the name of the relevant article and journal or book where it was published. Understandably, it would be best to give the author who discovered the fact or was the first to give a basis for it. However, in practice this is usually far from the case. Authors of articles usually cite the sources from which they themselves learned the given fact. Of course, at least theoretically it should be possible to follow the chain of references in older and older journals back to the original source. Scientific workers are glad when they are cited in the works of other authors. The purpose of a number of citations in scientific articles is thus to please (or corrupt) the relevant colleagues, who could well be amongst the reviewers of the particular article and thus decide on its acceptance for publication (see Box 2.7 on p. 19) or, at the very least, in the future can, in return, cite our articles in their works.

(including a very old and most probably outdated work citing 80% frequency of occurrence of toxoplasmosis in the inhabitants of Paris).

It is known about this parasite that it reduces the abilities of infected intermediate hosts (here should be placed, on the one hand, a citation demonstrating the reduced abilities of infected laboratory rodents and, on the other hand, citations of one of our earlier works on prolonged reaction times in infected persons, and one of our works showing 2.6-fold higher probability of a traffic accident in persons suffering from toxoplasmosis[6]). At the end of the introduction, it should be emphasized that, for understanding the importance of the Rh factor, nothing is more important than to perform reaction-time tests on Rh positive and Rh negative persons who are healthy and infected with toxoplasmosis and to determine whether, in accordance with our original hypothesis, Rh positive and Rh negative persons will react differently to infection. A sufficiently brazen author would, in addition, write (and certainly somehow justify) why voluntary blood donors are an especially suitable model group.

There is thus no doubt that such an "Introductory" chapter can be written only after completion of the Results chapter, in which we describe what we actually determined in our study and what we can allow ourselves to publish, and the Discussion chapter, in which we clarify what our results mean or, more exactly, how we will explain them to readers in our article. I certainly don't need to explain to the experienced person that this need not always be the same thing. For the inexperienced – in order to be able to determine how things really are in the future, we must first ensure that we will be able to do science at all. This is connected with the need to obtain money for our projects and thus to produce results through which we can demonstrate our scientific productivity to institutions providing funds for science and also to our employer. The results that are used to measure the productivity of scientific work do not consist in the discoveries we make, but the number of publications that we get into the press and also the number of citations of these works in the articles of our colleagues. The real importance of a scientific discovery is usually revealed many years later, while it is necessary for a great many reasons to perform regular evaluation of the productivity of scientific workers and scientific teams.

Thus, for example, if, during evaluation of the data, we find that men and women react differently to infection by toxoplasmosis, which does not in any way negate our hypothesis of the effect of toxoplasmosis on the current occurrence of Rh negative and Rh positive persons in the population, but makes it more complicated, we must carefully consider whether the first article should contain both the results for men and the results for women. The reviewers and editors of scientific journals who decide whether our article will be published or not generally don't like very complicated hypotheses.

Box 2.7 Review process in a scientific journal

If a scientific worker makes a discovery (and, in fact, also if he doesn't make a discovery), he must write an article about his results for a scientific journal. He sends the manuscript of the article to the editor of the journal and he then usually sends it to two or three reviewers, i.e. scientists who work in the same field and, where possible, in the same or a similar area. These are frequently members of the editorial board of the particular journal, whose or authors results or theories are mentioned in the article (especially if they are mentioned in a negative context) or authors who have published an article on a similar subject in the particular journal in the past. These reviewers (unless they happen to be your acquaintances who support you or who will require your favour in the future) attempt to find mistakes in the article that would form a basis for rejecting it. If no important mistakes are found in the article but they still don't like something about the results (for example, that they didn't discover them themselves), they think up some inadequacies (the author doesn't sufficiently discuss the possibility that ..., instead of method xy it would have been better to use method yx) and suggest to the editor that the article be rejected or at least be fundamentally rewritten (which, under current conditions with an excess of manuscripts of articles, is generally the same thing in the last analysis). On the other hand, if they like the article, find you empathetic or if it is useful for them if your article is published (for example, because they can refer to it in their works or because you cite their article in it in a favourable context), they recommend to the editor that your article be published. In any case, the final decision on the fate of the article lies with the editor who can, but need not, follow the recommendations of reviewers. Reviewers should be unknown to you; in actual fact, in at least half of cases, it is possible to guess who was involved. Especially in the case of favourable reviews, their authors usually take care that you will be able to guess their identities. In some journals, the reviewers do not obtain information from the editor on who is the author of the particular article; in others, the reviewer must sign his review. Studies that have been performed have, however, shown that this has minimal impact on the quality of reviews. It has been found that young reviewers and reviewers who are conversant with statistics write somewhat better

This ends my somewhat extensive description of a case from my laboratory and I can return to Darwin. (Who said "High time, too!"?)

How Darwin pulled the reader's leg

In the Introduction to his fundamental work "On the Origin of the Species by means of Natural Selection", Darwin writes that the main reason for his interest in studying the origin and development of the species and thus the main reason for formulation of the theory of biological evolution consisted in some facts related to the distribution of plants and animals and also the similarity of modern and extinct fauna in America, which he noticed when he travelled around South America as a scientist on the research ship Beagle. From the standpoint of "scientific marketing" such a starting point as the introduction to a scientific work is very correct. This was especially true in the middle of the 19th century when the official methodology of science was based on Bacon's empiricism, positivism and the inductive method.

Box 2.8 Empiricism, positivism and the inductive method

It would be a long tale to tell. Interested persons are referred to a dictionary of philosophy. Here, it is certainly sufficient to state that, according to the ideas of the philosophy of science in the 19th century, it is necessary in science to base our work purely on observed and measured data and to derive general rules directly from this data (i.e. from individual facts).

Irregardless of the methodical procedures employed by scientists of the time, they externally held to the basic principle of the inductive method, i.e. first we must collect scientific data, without any preliminary hypothesis as to how things should appear and what this should mean, and only subsequently, on the basis of evaluation of this collected data, can we form a hypothesis explaining the character of the obtained data and the relevant process that is responsible for the nature of the data. At his time, Darwin was one of the few scientists who opposed this dictate as the "only proper scientific procedure" and, in fact, openly defended the opposite approach – basically the modern **hypothetical-deductive Popper method.**

Box 2.9 Sir Karl Raimund Popper (1902–1994)

Popper was probably the most important philosopher of science of the 20th century. For example, he was concerned with the aspect of confirmability (verifiability) and refutability (falsifiability) of scientific theories (see also Box 3.8 on p. 46). It is interesting for evolutionary biology that he basically never understood it and simultaneously spoke about it very authoritatively.

Czech people will find it of interest that he died almost immediately after ancient and famous Charles University awarded him an honorary doctorate. In fact, it seems that an honorary doctorate or award from my alma mater is one of the most dangerous events that a person can encounter. It is surprising that the right to award prizes and honorary doctorates of Charles University has not yet become the subject of strict international control. Purely at random, the political map of the Near East could look entirely differently if someone had warned Shah Mohammad Reza Pahlavi against the danger of accepting an honorary doctorate from Charles University.[7]

In this method, a hypothesis is first formed on how a certain phenomenon could function, what its nature could be. On this basis, it is derived how certain data should appear if it were valid and if it were not valid, and only afterwards are the conjectured data capable of deciding on the validity or lack of validity of the given hypothesis collected. Darwin himself wrote that collection of data without any prior idea of how things could finally appear and what they could decide is a similarly ridiculous activity as setting out for the beach and sorting stones as to their size and colour. Nonetheless, in the introduction to his famous book, he kept to the traditional approach and, at least externally, pretended that he used the accepted inductive method. Simultaneously, it is almost certain that he arrived at his theory of evolution by a completely different approach – using the hypothetical deductive method. In any case, he could do nothing else. Although the material that he brought back from his famous expedition to the shores of South America contained informational evidence for the theory of the evolutionary origin of the species, I would guess that this information was not obvious enough to convince scientists unbiasedly observing nature of the existence of evolution.

Let's take, for example, **Darwin's famous finches**, given in all textbooks as an excellent example of evolution. This was a group of closely related species of buntings living on the Galapagos Islands. A similar, and in fact better, example consists in the Drepanididae species living on the Hawaiian Islands which, for a change, is a group of bird species that are very close relatives of finches. It is a pity that they did not receive the name **Wallace buntings** (after the second, independent discoverer of the theory of evolution). At least here, I can remedy this unpardonable error. It is typical for buntings from the group of Darwin's finches and for finches from the group of Wallace's buntings that they form groups of very close relatives, but have the life styles and thus body structures of quite distinct species. Some species have specialized in collecting small seeds, others in cracking hard seeds, others in catching insects and others, for example, in digging insect larvae out from under the bark of trees and from wood. The individual species thus occupied practically all the **ecological niches** on their group of islands that are occupied on the mainland by quite different species of birds from mutually unrelated groups.

Box 2.10 Ecological niche

A niche is a simple term for the life style of a particular species, the manner in which it utilizes the resources in its environment (sources of food and also shelter from predators), how much it is harmed by the individual physical, chemical or biological factors of the environment. The niches of various species can partly overlap; however, two species that have completely overlapping niches cannot survive for long in the same place. In addition, there is a negligibly small chance that two species will have exactly the same niches. Ecologists are divided into two groups with different opinions. Part of them state that a niche is created by species and that the term **"empty niche"** doesn't have any meaning. Another group of ecologists are of the opinion that an empty (unoccupied) niche is a logically incorrect term but simultaneously easy to understand intuitively and highly necessary.

The close relationship of species occurring on a single group of islands indicates that the individual species are not created independently of one another, but always from one another. If God had created the species or if they had perhaps been created independently of one another by autogenesis of nonliving material, some hawfinches cracking hard nuts, chickadees eating small insects and woodpeckers packing larvae out of wood would probably occur similarly on both groups of islands as on the mainland, or the relevant niche would have remained unoccupied. However, if species are created by biological evolution and the raw material for the creation of a new species would always consist in a species that occurs in the given place, it is quite logical that species on a particular group of islands will be mutually related and that only secondarily would they adjust with greater or lesser success to the individual means of subsistence. The rapidly progressing process of diversification of descendants of the original species is termed **adaptive radiation** by evolutionary biologists, see also Box 14.2 on p. 163. Which species becomes the "forefather" of all the other species on the particular group of islands is more or less a matter of chance. This will probably be the species that was the first to colonize the particular island, usually from the closest mainland. Understandably, if a particular group of islands were located close to the mainland or were even part of it in the past, in all probability the individual niches would be occupied by the same, mutually unrelated species of birds as on the mainland. Of course, adaptive radiation of species on groups of islands or individual islands does not apply only to birds, but can be encountered for a great many other groups of fauna and flora.

The samples of natural material, including finches from the Galapagos, that Darwin sent back to England during the expedition were extensive, but were certainly not ideal material for biogeographic studies permitting distinguishing of relationships between specimens of fauna and flora in adjacent territories and also for revealing the existence of the actual phenomenon of adaptive radiation on the islands.

Box 2.11 Biogeography

Biogeography is a science that is concerned with the study of the laws (and specificities) of the distribution of the individual species of organisms and the individual groups of organisms on the Earth. The presence or absence of species in a certain territory is explained in terms of differences in local conditions, the manner of migration (relocation in space from generation to generation) of the members of the individual species, changes in the spatial distribution of land on the globe and adaptive radiation of species at a particular site.

Darwin himself probably expected that the same kind of finches would be present on all the visited islands (he just didn't manage to find them all on every island), so that he mostly did not even state the island from which the collection item came. A pile of dried dead bodies became valuable scientific material useful for biogeography and later evolutionary studies only when they got into the hands of the appropriate expert in England, who classified them, determined the species and compared them with material obtained from other parts of the world.

Darwin's experience that he gained in collecting **fossils on the South American continent** could have been only slightly more useful. It is true that the fossils found included the remains of species related to species living in South America at the present time. However, I am not entirely sure that a non-palaeontologist could have recognized this from unprocessed material.

In fact, without any stratigraphic data, i.e. without knowledge of how the numbers of the individual species changed in neighbouring layers of the palaeontological record, and especially

Box 2.12 Palaeontology and palaeontologists

In contrast to evolutionary biology, which is concerned with the general laws of the development of life, palaeontology is concerned with the specific history of the alternation of species on the Earth. The main source of palaeontological knowledge consists in fossils, ancient remains of the bodies of organisms (or rather their hard parts) or remains left by their activities (paths, faeces) that have escaped decomposition by happy circumstances and have remained in better or worse preserved form to the present day.

without any information on the progress of exchange of species between North and South America following the formation of the Isthmus of Panama[8], Darwin could, at very best, only guess which of the living species currently occurring in South America actually belonged there and which wandered in during the recent past. It can again be concluded that the palaeontological data contains innumerable examples for the correctness of the theory of evolution; unprocessed palaeontological data, though, that a palaeontologist can encounter through visiting several palaeontological sites are not very useful in this respect.

However, let's stop guessing. Darwin kept very detailed diaries during his expedition. Neither in these diaries, nor in the extensive correspondence from that time is there even the slightest mention of anything that would indicate that Darwin in any way considered the transformation of species during the expedition. If the South American expedition contributed in any way to the discovery of the theory of evolution, then this was through provision of sufficient time for thinking things over and in that it allowed him to gain a reputation as an important natural scientist. Especially the latter was truly invaluable for Darwin, who was a Bachelor of Theology and did not have any official education in the natural sciences. Diaries and letters from a later period indicate that Darwin arrived at his theory of evolution only after returning from the expedition, sometime between 1836 and 1838, and that the stimulus for its creation probably came from completely different sources.[9] Only after he thought up the theory of evolution (which, however, he himself, called not the theory of evolution but the theory of the origin of species), did he look back at his collections and his notes from the expedition to determine whether they could support his theoretical conclusions.

What Darwin actually discovered and how

How did Darwin actually arrive at his theory and what was the nucleus of his discovery? Surprisingly, not in the discovery of evolution. In the 19th century, the possibility of evolution, gradual development and inter-conversion of the individual species was the subject, not only of scientific hypotheses, but also of successful popular scientific books. Fifteen years before the publication of Darwin's book "On the Origin of the Species by means of Natural Selection", Robert Chambert anonymously published the popular work "Vestiges of the Natural History of Creation" (1844), in which he discussed the possibility of evolution explicitly and in great detail. This book came out in eleven editions by 1860 and a total of 24 000 copies were sold in the ten years following the first edition.[10] For comparison, Darwin's book "On the Origin of the Species by means of Natural Selection" sold only 9 500 copies in the first ten years. The aspect of the existence and possibility of the formation of adaptive traits in living organisms constituted the subject of a number of theological works with which Darwin must have been acquainted during his study of theology. However, what was lacking at that time was a scientific hypothesis offering a satisfactory explanation for the driving force for the development of species, their mutual diversification and especially what causes the emergence of adaptive traits, i.e. the formation of organs and patterns of behaviour effectively assisting survival of the organism under its natural conditions.

And this was the main contribution of the theory of evolution of Charles Darwin. The natural selection that he discovered simultaneously offered an answer to two important questions. He explained why species change and simultaneously convincingly explained **why they exhibit adaptive traits**.

Box 2.13 Adaptive traits

Organisms exhibit a vast number of properties (organs and patterns of behaviour) that assist in their successful survival and reproduction. Some organs assisting in successful survival are quite simple and their usefulness and means of evolution are easy to discover (fins for swimming, parachutes on dandelion seeds for dispersion), while others are highly ingenious. For example, some kinds of orchids (*Ophrys*) have a structure in their flowers whose shape, colour and scent are similar to the females of a certain kind of fly. Thus, males are attracted to the flowers and attempt to copulate with the dummy female and thus transfer pollen from one flower to the next. Tobacco and cotton plants can recognize that they are being eaten by the caterpillars of the moth *Heliothis virescens* (tobacco budworm) (they can even distinguish that they are being damaged by this pest and not the caterpillars of some other kind of moth or a scientist punching holes in the leaves). In order to get rid of the intruder (or at least to make his life harder), they begin to emit chemical substances that attract the natural enemies of this kind of caterpillar, the parasitoid wasp *Cardiochiles nigriceps*, which lays eggs in the caterpillar. These parasitoid wasps fly to the plants even if the scientist first removes the caterpillars and the damaged leaf. The plants do, of course, not know that they are doing this – in this sense, it is not truly goal-oriented behaviour. However, it is certainly useful behaviour as it truly effectively assists the plant to get rid of the particular species of pest.[11]

The South American natural science observations that Darwin mentions in the first chapter of his book could not have substantially assisted in the discovery of natural selection. However, the knowledge that he gained as an enthusiastic pigeon breeder in his new home in the small village of Downe (Kent) in England could have contributed far more. Pigeons, similar to dogs, are excellent and extremely malleable material in the hands of a breeder. A good observer, which Darwin undoubtedly was, basically could not avoid coming to the conclusion, on the basis of experience with this species, that species are fundamentally variable and that, after being subjected to the appropriate selective pressure, will change over time in any way the breeder wishes.

Of course, artificial selection, which is responsible for the development of mutually dissimilar strains of pigeons or varieties of useful plants, presumes the existence of a breeder, who

Box 2.14 Selective pressure

Selective pressure is pressure exerted by the environment or man on a certain population through removal of the bearer of certain traits, e.g. an above-average large or below-average small individual, or by preventing such an individual from reproducing. Selective pressure need not always mean a negative effect on the bearer of undesirable forms of the trait, but can just as well consist in support for individuals with the desirable form of the trait.

purposefully decides which individuals will be able to reproduce and pass their properties on through their descendants to future generations. However, such a breeder is lacking in nature, at least if we decide not to consider intervention on the part of a God. He would most probably use more effective methods of creation of new species and, at the very least, would not spend the unbelievably long period of time (in fact, 3.6 billion years) doing this.[12] And here is the real brilliance of Darwin's discovery. After he became acquainted with the work of **Thomas R. Malthus** (1766–1834)[13], Darwin realized that this breeder, who constantly decides which individual will transfer its traits to another generation, is nature itself. Malthus basically discovered that an intra-species battle is constantly waged in nature for resources. He came to this unexpected conclusion because he took a very atypical species – man – for his considerations. While the populations of other fauna (and flora) remain stable as to numbers in the long term, or more or less oscillate around a certain value, the human population has been growing constantly (at least in living memory). Today, we would say that, on the basis of known information and on the basis of a theoretical model, Malthus demonstrated that this growth must necessarily be **exponential**, i.e. that it constantly accelerates with time.

Box 2.15 Exponential and linear growth

If the population in each generation increases by a constant multiple and if, for example, it doubles in each generation, this is called exponential growth. Exponential growth is constantly faster – if there are ten individuals in the first generation, there will be twenty in the second, forty in the third, eighty in the fourth, etc. In contrast, linear growth occurs when the number increases by a constant amount in each generation, e.g. by 10 individuals. Linear growth occurs at a constant rate – if there are ten individuals in the first generation, there will be twenty in the second, thirty in the third, forty in the fourth, etc.

If the amount of available resources remains constant or increases only linearly, i.e. grows at a constant rate, there must necessarily come a time when the resources start to become inadequate and human beings begin to experience famines. I have no idea how Malthus came to the conclusion that the amount of resources available to humans would increase linearly. I am afraid that he did not base this on any real data or on a theoretical model and that he was in all probability wrong on this point. However, this is not important from the standpoint of the evolutionary importance of his work. What is important is that his famous essay inspired Darwin to consider the disproportion between the large reproductive ability of any population and the constant number of its individuals.

These considerations finally brought him to the conclusion that the constant lack of resources to which all species will work their way sooner or later leads all species to a point where superfluous individuals, for whom there are insufficient resources in the environment, are gradually eliminated from the population. This constant **removal of superfluous individuals** is not a random process. There is greater probability that those individuals whose traits correspond least to the requirements of their environment will be removed from the population. This provides an automatic advantage to individuals who are best adapted to the given requirements. Thus, nature acts as a tireless breeder who, from the beginning of time, consistently selects from the population those individuals whose accidental deviations from the usual shape or usual traits (from the standard) were found to be advantageous in the fight for resources and allows these individuals to preferentially reproduce. Simultaneously, the traits of parents are very frequently transferred to their descendants, i.e. are inherited from one generation to the next. As differences in the traits of individual organisms (deviations from the standard) appear again and again in each generation, and in all possible directions each time, species must constantly change over time. And that is not all. Because nature, as the breeder, is constantly making a decision from the same point of view, i.e. according to the ability of individuals to better utilize the resources in the environment for survival and reproduction, the adaptation of organisms to the natural conditions must get better and better, i.e. **adaptive traits must be accumulated**.[14] While pigeon breeders create new breeds according to their ever-changing preferences and thus obtain breeds with the most varied, frequently strange traits, nature subjects its selection program to a single target – increasing the ability of organisms to obtain resources from their environment and to use them for the production of offspring. The most obvious and simultaneously the most mysterious property of living organisms, i.e. the presence of adaptive traits, is a necessary consequence of this "targeted" selection program.

The brilliance and main reason for the success of Darwin's discovery consists in the fact that he was the first to bring together knowledge related to several long-known processes in a logical whole. He demonstrated that the result of these generally recognized processes must be natural selection and evolution of species leading to the formation of adaptive traits.

It is not difficult to convince any sensible person that the initial assumptions of Darwin's theory of evolution are reasonable, whether he knows anything about biology or not. It is relatively obvious that two individuals produce an average of more than two offspring during their lives and thus that the population should grow over time. It is also apparent that this does not happen and that the populations of most species remain more or less constant, so that excess individuals must be removed from the population or at least not reproduce during their lives. It is also more or less evident that all the individuals of a single species are not completely identical and that they differ in a great many traits. It would also seem quite obvious that, amongst individuals with different traits, those whose traits are better adapted to the conditions in which they live, i.e. those who are better able to obtain resources under these conditions and use them for production of offspring, have a better chance of survival and reproduction. And everyone who has looked at his progeny or at parents and children in his surroundings can see that traits are inherited by offspring from their parents.

A propos, **comparison of the similarity of parents and children** is probably an especially suitable way of convincing oneself that physical traits are actually inherited. In fact, this could even be a method that is too good. Let me explain. The results of some studies have indicated that the similarity that we perceive between fathers and their very young (however, not their older) children is suspiciously high. Some authors are thus of the opinion that, in relation to the similarity of parents and children in humans, this could be an adaptive trait of our species whose biological function is to prevent **infanticide** (killing offspring) by fathers. "Mother, don't tell me that that screaming infant is mine. Hand me a club." "But, father, look at that flat nose, the blue eyes and the aristocratic eyebrows. That's exactly how you look." "Mother, I could believe you, but I don't have to, don't forget that we haven't invented the mirror yet." "Aw, come on, father, ask your own mother, or remember what our friends always say when they come to our cave for a visit – nu, nu, nu, nu, he looks just like his daddy."[15]

If individuals truly have, on an average, more offspring than live to see adulthood, if individuals of a single species differ from one another, where the probability of surviving to adulthood and reproducing depends on their traits, and if traits are truly inherited by children from their parents, then evolution of the species and the formation of adaptive traits must occur. The greatest mystery was suddenly why mankind came upon such a simple and correct explanation only in the middle of the 19th century. Even this question is not difficult to answer. It wasn't until the 19th century, at the time of developing capitalism, that Europeans could watch the work of natural selection more or less as a live show and feel its effects more or less on themselves. Evolution proceeds very slowly in nature and is mostly not very apparent. Consequently, even when logical considerations lead us to the conclusion that evolution must necessarily occur and species must develop over time, our experience will constantly convince us otherwise. It seems like a horse has been a horse, a rabbit a rabbit, and a pine tree a pine tree since the beginning of time and can in no way be seen to change from one generation to the next. In contrast, spontaneous development occurs in society, especially at a material level. In the 19th century, this development became so fast that it was very difficult to overlook. New inventions appeared with increasing regularity and entered the lives of ordinary people with increasing speed. New companies were established, competed with one another and the less successful, producing worse or more expensive products, disappeared, while those that produced better products were successful and grew. Simply, an excellent parallel to natural selection. And if something like this could function in good old England, why not believe that the same process could occur in the world of plants and animals.[16]

Why were biologists (in contrast to the general public) unhappy with the theory of evolution

At the meeting of the Linnean Society of London at which Darwin's and Wallace's theories of evolution[17] were first present to the public in 1858, the results of the two scientists did not make a great impression. In his final speech, the chairman complained that no unusual discoveries had been made that year. However, the new theory fell on fertile soil amongst the general public. Over the next few years, Darwin's book was translated into all major languages and his theory spread successfully around the world. Understandably, Darwin's concepts met opposition amongst a number of people. This opposition tended to be at an ideological level and was directed mainly against the possibility of using the conclusions of the theory of evolution to explain the creation and development of man. I am personally of the opinion that Darwinism was spread best through the discordance of Darwin's concepts of the origin of man from an animal forefather with the concepts declared by the church and in Sunday School. For a great many people, the bringing of man down from his exclusive position to the earth amongst the other creatures, for others the reduction of religious authority and the role of God in controlling the world made Darwin's theory an interesting subject of social conversation, at the very least. All these facts meant that Darwin's theory was far more favourably received than could have been expected for such a fundamentally new and important theory.

However, the spreading acceptance of Darwinism was resisted for a long time by a relatively small, but quite important group of people. Surprisingly, these were scientists who were professionally concerned with this subject at that time. Understandably, the reason for this could be ordinary human malice and jealousy. It is not easy for anyone to admit that a newcomer to the field, who emerged as if from nowhere, could arrive at the correct (and what is more, simple) explanation of a problem on which one has intensely and unsuccessfully worked for years. However, if we look closely at the most important objections of Darwin's opponents from amongst the professional public, we can readily see that psychological causes were not the only reasons for rejecting Darwin's theory.[18] To be on the safe side, let's repeat the initial assumptions for the validity of Darwin's theory of the origin of the species by means of natural selection:

1. On an average, an organism has more than one offspring during its lifetime (or a pair of parents more than two offspring) – thus the population should grow.
2. At the same time, the number of individuals remains constant in the long term – the excess offspring die without reproducing.
3. Individuals of a particular species differ from one another (i.e. there is variability within the species).
4. The probability of surviving to adulthood and reproducing depend on the traits of each individual.
5. The traits of individuals are inherited – on an average, descendants are similar to their parents more than to the other members of the population.

Darwin stated that: if points 1–5 are valid, biological evolution must necessarily occur, species must change over time and adaptive traits must accumulate.

In contrast to the general public, professionals very rapidly discovered that there are some very serious inadequacies in Darwin's seemingly iron logic. Scientific opponents raised the following objections: "Even if points 1–5 are valid, biological evolution need not occur; this depends on the value of some parameters of organisms. In real organisms, these parameters attain such values that evolution cannot occur."

Where is the hitch according to the opponents? According to Darwin's contemporaries and a great many of their successors, particularly points 3 and 5 are contentious. The very existence of variability within a population, i.e. differences between individual members of a particular species, is not sufficient for the functioning of evolution; it also depends **how much** of this variability exists in the population and **how fast** it is generated.

Box 2.16 Variability within a population

This is more or less the same as polymorphism in a population or, simply, heterogeneity. The members of a single species (to be exact, it should be added – of the same sex and age) differ from one another in external appearance and internal traits. If we simplify this a bit, we can state that the different effects of the environment, for example, different nutrition, are responsible for some differences (and these differences are not transferred to offspring), while other differences lie in the genotypes of the individual organisms (and these differences are inherited from their parents by descendants).

For example, imagine that, within a single species, individuals could differ in only ten traits and that the individual traits could assume only three values (small, medium, large). In such a case, natural selection would rapidly use up the variability present in the population, some traits would disappear from the population and others would become fixed, i.e. would spread to such a degree that they would be exhibited by all the individuals in the population. Even if variability is constantly generated in the population, this is of no help to evolution if there are only a limited number of ever-repeated variants. "Aw, come on!" Darwin would protest. "Let's not try to bring the long-dead ideas of the essentialists back to life."

Box 2.17 Essentialism

This is a branch of philosophy that assumes that the observed properties of actual objects are only more or less perfect manifestations of ideal internal properties, their nature, or essence. The concept that essential ideal properties do not actually exist and that there are only the imperfect properties of actual objects is the opposite of essentialism. In this approach, essential properties, such as roundness or redness, are only the product of human thinking arising from generalization and naming the observed properties of real objects.

"Do I have to remind you (Darwin is still speaking) that there are practically an infinite number of traits, i.e. the individual properties of organisms, and that they are mostly continuous in character, i.e. there is no reason to suppose that they could assume only a finite number of states. If an elephant has an average trunk length of 1.5 metres, sorry, I meant one fathom, there is no reason to infer that its offspring could have a trunk 1.1 fathoms long or, for all I care, four fathoms."

Okay, let's say that Darwin is right and let's take it for granted that the traits that can be exhibited by an organism are, in fact, infinite in number, that they can assume any values and that new variants are constantly formed in the population. Even then, it isn't clear that such variability in the population can be useful raw material for Darwinist evolution. It is an essential condition for the functioning of this evolution that variability must be formed in the population with at least the same speed as it disappears. And a great many of Darwin's opponents thought this is not true. Most species that we encounter in nature reproduce sexually. Each individual of a sexually reproducing species has two parents and the properties that it inherits from them are mostly more or less an average of the properties of these two parents. If an individual has a tall father and short mother, his size will be somewhere between them, closer to the average. If he has a large mother and an average-sized father, his size will again be somewhere between them, i.e. above-average, but not to the same degree as his mother's size. This **averaging of traits** has, however, the disadvantage that it rapidly removes variability from the population. In fact, it can be derived mathematically that half the variability present disappears in each generation.[19] Such a rate of decrease in variability is unacceptable from the standpoint of the possibility of functioning of Darwinist evolution. It would have to be balanced by the creation of variability with at least the same speed. However, such fast creation of variability is not encountered in nature; deviations from the average in the population are formed at a much smaller rate. In fact, the formation of variability at too great a speed would prevent functioning of evolution as it would practically exclude the heredity of traits. The appearance of an individual would be determined by the properties newly formed in him and not the properties that he inherited from his parents.

H.C. Fleeming Jenkin (1833–1885) offered a very evocative, although from a modern point of view not very politically correct, example illustrating this weak aspect of Darwin's theory of evolution. Imagine that a white man is shipwrecked on a tropical island inhabited by only black men and women. Because of his excellent psychological and physical qualities (in all probability it was an English gentleman) he rapidly achieves the position of head of the tribe. He wins all the fights and tribal conflicts and, before he dies of general debility in his old age, manages to have a large number of children with his own and other women. The problem is that the children are only half his and thus they are only half as white and good as he is. In subsequent generations, these children cross with the local population, become blacker and ever more ordinary, until they recall the original valiant shipwrecked man only through legends.

Darwinism fights back – the fitting (but well concealed) answer of the Brno abbot

Darwin took the objections of his opponents very seriously. In later editions he gradually modified and supplemented the text of the Origin of the Species so that the originally simple theory became ever more complicated and tangled. In the last edition, Darwin even doubted the importance of natural selection for the process of evolution and admitted a number of other possibilities, including **Lamarcks's concepts** of strengthening the organs by their use and inheritance of these acquired traits from generation to generation.

Box 2.18 Lamarkian model of evolution

This model of evolution assumes that the adaptive traits of modern organisms are formed in that the members of a certain species begin to devote themselves to certain activities, for example, they reach the tops of trees for leaves, in this "exercise" they lengthen their necks and their offspring then inherit these prolonged necks. It cannot now be determined whether Lamarck really had such a naïve idea, he did not state things so explicitly in his work "Philosophia Zoologica". That is, however, not important today – Lamarckism is now understood as the formation of adaptive traits through the relevant "exercising" and subsequent inheritance of these acquired properties.

However, this abandoning of the position was premature and unnecessary. An appropriate answer to the most fundamental objections of opponents of Darwin's theory of evolution was published in 1865 by the Augustinian monk and abbot of the monastery in Brno, **Johan Gregor Mendel**. We can only guess today what the original target of his study of heredity of traits in peas was. Glancing through his carefully underlined and densely annotated copy of "The Origin of the Species" however indicates that he was concerned with aspects of evolution and that his studies were most probably related to this area. If this conclusion is correct, then it is highly probably that, in his experiments, he attempted not to confirm Darwin's theory, but to overturn it.[20] Mendel's experiments showed that, although it does not seem so superficially, the predisposition for certain traits are passed on from one generation to the next in unaltered form. The progeny of peas with red and white flowers do have pink flowers; **the predisposition for red and white flowers is, however, in no way altered in the cross varieties.** We can verify this if these cross-varieties, called cross varieties of the first filial generation (denoted F1) are again crossed together, obtaining crosses of the second filial generation (denoted F2). Because the F1 crosses have, in their cells, one variant of the gene, shortened by one allele, from a red parent and one from a white

Box 2.19 Box 2.19 Allele

A variant of a gene, differing from other variants of this gene in its manifestation, is called an allele. One variant of the gene can be responsible for brown eyes and another for blue eyes. I would probably manage to write the book without the term "allele", and could use the term "gene variant" instead. But my colleagues who might happen to read the book would laugh at me, saying that I have lost my memory for even the most basic genetic terminology (and they wouldn't be that far from the truth), or they would not be sure whether I were speaking of alleles or of something else. So I guess you have no choice but to get used to the term "allele" (and the term "phenotype", see Box 2.22). At least at the beginning, I will give both possibilities (allele and gene variant), so I am sure we will manage it. After reading the book, you can surprise your friends with your newly acquired knowledge (or throw them off balance – which is also okay).

parent, and because each of their progeny obtain one copy of the gene from each parent, approximately one quarter of the crosses of the F2 generation inherit two alleles for red flowers (they will have the rr genotype) and will thus be red, approximately one quarter of the progeny will inherit two alleles for white flowers (ww genotype) and will thus be white and the remainder, thus approximately half the progeny will inherit one allele for red flowers and one allele for white flowers (rw genotype) and thus will be pink like their parents (Fig. 2.2).

Nothing actually happens to the alleles (gene variants). It makes no difference whether crosses of the F1 generation sometimes remind us of only one of the parents (see Box 10.1 on p. 131) – the nature of the two alleles and their frequency in the population do not basically change from one generation to the next (assuming that no role is played by selection, which would remove the carriers of certain alleles from the population).

Box 2.20 Genotype

Genotype is a combination of alleles (gene variants) borne by a specific individual in his cells (cell – see Box 5.2). In diploid organisms, each individual has a pair of alleles from each gene in his cells, where this can be a pair of identical alleles (**homozygote**) or a pair of different alleles (**heterozygote**). Because the number of genes in the genome (see Box 3.3) of organisms is enormous and a large percent of them occur in many variants in a given species, the number of possible combinations of alleles – number of different genotypes – is unimaginably large and practically no pair of individuals in the population of a sexually reproducing species, with the exception of identical twins, has an identical genotype.

It might seem at first glance (and I am convinced that Mendel understood the results of his experiments in this way) that these conclusions overturn the Darwinist model of evolution. Where would any evolution occur if the predisposition responsible for the relevant trait is invariable in time? In actual fact, the opposite is true and **hard heredity**, which was first demonstrated by Mendel in his experiments on peas, is an essential precondition for the functioning of Darwinist evolution.

The original concepts of the nature of heredity, based on observation and intuition, assumed that predispositions derived from the two parents affect one another and are averaged in the

Box 2.21 Hard and soft heredity

Hard heredity consists in the transfer of the predispositions for individual traits from one generation to the next in unaltered form, without any effect on the other predispositions present in a particular individual and the effects of the external environment. In contrast, soft heredity assumes that predispositions can change from one generation to the next under the effect of the other predispositions present in a given individual and through the effect of the external environment. The Lamarckist theory of evolution and the later Darwin's theory of evolution are based on the concept of soft heredity of predispositions; in contrast, the Neodarwinist theory of evolution, i.e. the main direction of the theory of evolutionary biology developed roughly from the 1930s and thus including the knowledge of Mendelian genetics, is based on the concepts of hard heredity.

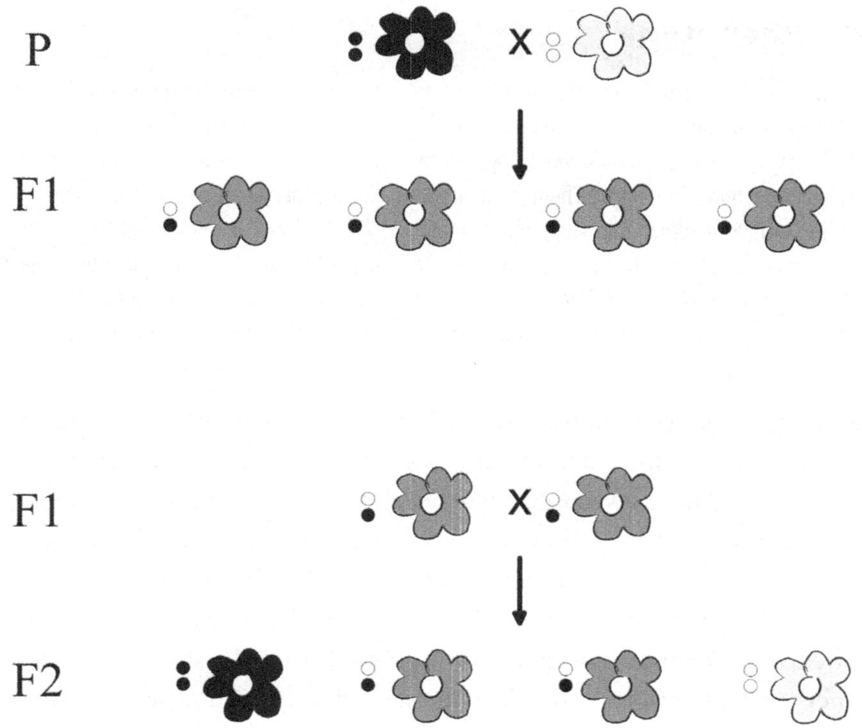

Fig. 2.2 Scheme of inheritance of traits with incomplete dominance. The upper part of the figure depicts the result of crossing two parent plants (denoted in genetics by letter P), red (black in our figure), carrying two alleles for red flowers, and white, carrying two alleles for white flowers (the alleles are denoted by dark and light circles). The crossing results in plants of the first filial generation (denoted F1), which have one of each of the alleles and are thus all genetically identical and have the same appearance (grey in our figure, in actual fact pink). The lower half of the figure depicts the result of crossing two individuals in the first filial generation. Their offspring (the second filial generation, F2) differ from one another genetically and in their appearance. One quarter have both alleles coding red flowers, one quarter have both alleles coding white flowers and one half has one of each allele (and are thus pink, like their parents).

bodies of crosses. Mendel's model of heredity, based on the results of his experiments, demonstrated to the contrary that this is not true and that the predispositions (today we know that these are the individual alleles) are transferred from one generation to the next in unaltered form. And where does evolution come from then? As we mentioned above, neither individual alleles nor their frequency in the population are changed as a result of crossing and the representation of the individual alleles in the population is also not changed as a consequence. However, there is a nonzero probability that alleles can be altered as a consequence of other processes, such as **mutation**, see Chapter 3. The number of copies of the individual alleles in the population– and thus the appearance, called the **phenotype** – of individuals in the population changes over time, e.g., because the individuals of various phenotypes have an advantage or disadvantage to various degrees through natural selection.

Box 2.22 Phenotype

This term denotes the combination of all the traits of an individual, including its behaviour. With a certain degree of simplification, it might be possible to use the word "appearance" instead of phenotype; however, it must be borne in mind that phenotype also includes internal properties, which are not visible externally, and also behaviour. Thus, it will probably be more practical to use the proper scientific term "phenotype". At least the reader will be each time reminded that this is a technical scientific term, for which there is an exact definition and which all scientists use (or at least should use) in the same way. This is, incidentally, the reason why scientists use scientific, frequently Latin or Greek, expressions and why they don't speak plain English. The meanings of words in normal language are not exactly defined and thus scientists could not discuss things unambiguously. "You said appearance, my colleague. Did you mean by this also the shape of the pancreas?"[21]

It is a great pity that Darwin was not acquainted with the results of Mendel's experiments or, at the very least, did not understand their actual importance for his theory of evolution. Perhaps this knowledge could have saved him a few sleepless nights. I am not sure whether it would have been a pity for Mendel. His quite fundamental contribution to the theory of evolution would probably have brought him recognition in the eyes of Darwin and other evolutionary biologists, but would definitely not have improved his position with his religious superiors. (While I have little knowledge of the running of church institutions, I suspect that evolutionary biologists would probably not – from their positions – be entrusted with the naming and recalling of abbots.)

Summary and incitement

At the end of this chapter, I will summarize things up so that it is clear that the core of the information did not lie in toxoplasmosis or how to write a scientific article. Darwin certainly did not discover evolution – the variability of species over time. His fundamental contribution to the creation of the scientific theory of evolution lay in discovery of the forces that can drive evolution, which can lead to the formation of adaptive traits, specifically the discovery of natural selection. However, at Darwin's time, it was assumed that the heredity of biological traits is soft, that predispositions for individual traits inherited from both parents are averaged in the progeny. Darwin's theory of evolution could not function under this assumption, as averaging would rapidly lead to disappearance of all genetic variability from the population. It was the discovery of Mendelian genetics that demonstrated that heredity is "hard", that only the effects of predispositions from both parents can be averaged in progeny, while mostly nothing happens to the actual predispositions that demonstrate that Darwin's theory could be correct. The next chapter will again mention the formation of Neodarwinism, i.e. the product of the synthesis of the Darwinism of the 19th century with the genetic and biological discoveries of the first half of the 20th century.

Footnotes

1. The slower reaction of people with latent toxoplasmosis was first described in an article in the journal *Parasitology* 122: 515–520, 2001. The results indicated that, in the first minute of the test, infected persons responded to simple instigations with the same speed as uninfected persons; however, in the second minute, the performance of infected persons deteriorates. In this work, the difference between the responses of Rh positive and Rh negative persons had not yet been discovered; subsequent analysis of the initial data showed that this effect is also manifested.
2. An interesting article on the manipulative activity of the parasite *Toxoplasma gondii* was published in 2001 by J.P. Webster, *Microbes and Infection*, 3: 1037–1045, 2001.
3. So how should a gene be properly defined? This is difficult. A gene is a predisposition responsible for one specific smallest independent definable difference between two individuals in the population. In some cases, a single DNA section can be responsible for this difference (to be more precise, one or more differences located in a certain DNA section); at other times, it is the difference between individuals determined by the sequence of the DNA at several places on the genome. A difference in a single nucleotide in the DNA can cause a change in the properties of an organism and can thus be the material substance of a gene. If a gene is defined on the basis of the difference between individuals, it is apparent that the particular property (e.g. eye colour) must exist in the population in at least two forms. A property that occurs in the population in at least two forms is termed a trait. We don't learn about the existence of a great many genes, as the relevant differences in the properties of organisms can appear only under certain situations (in combination with a particular effect of the environment or with certain alleles present in the particular individual), or variability of the particular gene can be lacking in the population (or even in the studied species) at the particular moment. The concept of molecular biologists that we can count the genes in a particular organism by scanning its genome is completely naïve – the number of genes understandably substantially exceeds the number of individual nucleotides in its genome.
4. It is a puzzle how both variants of the Rh gene can survive in the population and how the new variant of the Rh gene could appear in the population at all (until our discovery of the protective effect of the Rh factor against the detrimental consequences of toxoplasmosis). If we forget about toxoplasmosis, the carrier of the variant that is in the minority in the population is always at a disadvantage compared to the carriers of the more common variant and the new variant (e.g. the newly mutated variant of the Rh gene) can thus not successfully penetrate into the gene pool of the population. Imagine that nearly all the members of the population are Rh positive. Then, Rh negative women will have, on an average, fewer children than Rh positive women, as some of their children will die as a consequence of damage to the red blood cells by the antibodies of the mother. The same is valid for the opposite case. The Rh positive form of the gene cannot spread in a population consisting of Rh negative persons as, in this case, Rh positive men will have, on an average, fewer progeny than Rh negative men (once again, some of their children will die as a consequence of damage to the red blood cells).
5. In 2000, British authors published an article indicating that a rat infected with toxoplasmosis stops being afraid of the smell of cat urine. *Proceedings of the Royal Society of London Series B-Biological Sciences* 267: 1591–1594, 2000. Three studies of American and British authors yielded the same result: *Proceedings of the Royal Society B-Biological Sciences* 273: 1023–1030, 2006, *Proceedings of National Academy of Science* 104: 6442–6447, 2007; *Neuroscience* 148: 342–348, 2007.
6. Amongst persons injured in traffic accidents being treated in the emergency department of a hospital in the centre of Prague, i.e. injured drivers and pedestrians who were hit by a motor vehicle, we found a considerable difference in the number of persons with latent toxoplasmosis compared to a control sample of persons of the same age amongst the population in Prague. It can be calculated from the fraction of infected persons in the two samples that infected persons have, on an average, a 2.6-fold greater probability of being in an accident than healthy persons. The greater the antibody level in these persons (i.e. the stronger or fresher their infection), the greater the risk of an accident. If toxoplasmosis has similar manifestations in other parts of the world, it is quite possible that the number of persons who die each year from injuries as a consequence of toxoplasmosis approaches the number of persons who die from the worst protozoan parasite disease – malaria. *BMC Infectious Diseases* 2: 11, 2002. In 2005 an independent study performed in Turkey confirmed the effect of toxoplasmosis on the risk of an accident: *Forensic Science International* 163: 34–37, 2006. Recently, similar results were also obtained in our new large prospective study on 3 890 military drivers.

7. While western-oriented Shah Muhammed Reza Pahlavi received an award and honorary doctorate from Charles University in the Great Hall of the Prague Carolinum, the Islamic Revolution broke out at home in Iran and deposed him from the throne.
8. The flora and fauna on both American continents had developed in an isolated state for a very long time and thus differed drastically. The Isthmus of Panama was formed only 3–4 million years ago and a number of species crossed it in both directions. Especially the species originally from North America became very well established on the new continent and were apparently the cause of the extinction of a number of groups of South American species. However, newer studies have shown that North American species apparently did not win in the battle for resources, but in species selection. In contrast to South American species in North American, North American species in South America repeatedly speciated and thus replaced the species that arrived originally (which died out over time in both Americas).
9. I will not pretend that I have, myself, read Darwin's correspondence. However, if you are interested, this is possible (now on the Internet): *The life and letters of Charles Darwin*, including an autobiographical chapter. Ed. Francis Darwin. 3 parts, Murray, London, 1887, *More Letters of Charles Darwin. A Record of his Work in a Series of Hitherto Unpublished Letters.* Ed. Francis Darwin and A. C. Seward. Murray, London, 1903, *The Correspondence of Charles Darwin.* Vol. 3: 1844–1846. Cambridge Univ. Press (Cambridge), 1987. His travel diary is also available: *Charles Darwin's Beagle Diary.* Ed. R.D. Keynes. Cambridge University Press (Cambridge), 1988.
10. A highly successful book, *Vestiges of the Natural History of Creation*, was published anonymously by R. Chalmers in the John Churchill Publishing House (London) in 1844.
11. Experiments on plants attracting parasitoid wasps were described in the journal *Nature* 393: 570–571, 1998.
12. I must admit that this is entirely my subjective and basically unsupported opinion. God could, of course, mess around with creating organisms for any length of time and, as a certain Jewish joke points out, what may seem like eternity to us, could be no more than a second for God. In addition, God could certainly have chosen a slower way of creating life so that he did not clearly reveal his role in this process and so that we could (amongst other things) try to explain the creation of life by natural means. I don't know about you, but I would tend to accept his invitation in this respect.
13. Darwin apparently read the 16th edition of 1826, but the famous essay of T.R Malthus was originally published in the 18th century: *An essay on the Principle of Population, As It Affects the Future Improvement of Society.* J. Johnson (London), 1798.
14. Adaptive traits of organisms are very frequently confused with goal-oriented traits. However, there is a considerable difference between these two concepts, which is easiest to explain on specific examples. If we catch the flu, we can visit a physician or an exorciser. In both cases, this will be goal-oriented behaviour, because we are doing this to get rid of the disease. However, only the visit to the physician will be adaptive behaviour, because it can actually help us to get better. (However, if you have a bad physician and good exorciser, then who knows?) All organisms are characterized by the presence of a large number of adaptive traits. In contrast, goal-oriented behaviour has been demonstrated only in some species, such as man and chimpanzees.
15. A study on the suspicious similarity of small children and their fathers was published in 1995 in *Nature* 378, 669; objections related to the absence of mirrors in the caves of our predecessors were published a year later, *Nature* 379, 292, 1996. It is not entirely clear how nature can ensure the similarity of fathers and their small (but not larger) children. I would personally suggest that children defend themselves against infanticide by imitating the expressions on the faces of their parents.
16. Karl Marx was the first to point out that comparison with the development of early capitalist society is a good way of understanding evolution and he wrote about this observation in a letter to Engels on June 18, 1862. At the beginning of the 20th century, the same idea emerged quite independently in the works of other authors (amongst Czech authors, for example, Emanuel Rádl in 1908). In the second half of the 20th century, some authors extended this to the generalization that we are capable with only the greatest difficulties of distinguishing any rules and phenomena in nature if we haven't previously encountered them in some form in human society. This interesting phenomenon, called the sociomorphological model, is discussed, e.g., by Topitsch E. in *Ursprung und Ende der Metaphysik.* Springer (Wien), 1968, Peters H.M, in *Ratio* 1960: 22–37, 1960 and is described systematically by S. Komárek, e.g. in *Nature and Culture, The World of Phenomena and the World of Interpretations* (in Czech), *Vesmír* (Prague), 2000, and *Mimicry, Aposematism and Related Phenomena.* Lincom (Muenchen), 2003.
17. Darwin developed his theory in 1837–1838. In 1844, he completed its a brief description which, however,

he did not publish, but only gave to his wife with instructions to make it public in case of his death. In the following years, he devoted himself to collection of data to support his new theory and especially to creation of his professional reputation and network of social (professional) contacts. However, in 1858, he unexpectedly received a letter from the then-young A.R. Wallace, in which this biologist asked him to evaluate and possibly make public his own theory of evolution. Darwin was crushed, because he discovered that Wallace's theory was practically identical with his own. Finally, the situation was resolved in that, at a single meeting of the Linnean Society, extracts from Wallace's letter and from Darwin's original sketch of the theory of 1844 were read out. Darwin rapidly completed his long-delayed work *"On the Origin of the Species by Means of Natural Selection"* and published it in 1859.

18. You can read about the period opinions of professional critics of Darwin's theory, e.g. in the book by D.L. Hull: *Darwin and His Critics. The Reception of Darwin's Theory of Evolution by the Scientific Community.* The University of Chicago Press (Chicago), 1983.
19. Mathematic derivation of the disappearance of variability can be found in textbooks of population genetics or in the classical book of R.A. Fischer, *The Genetical Theory of Natural Selection.* Dover Publications (New York), 1958. In fact, this book contains (a rough outline of) all the interesting things that were only much later elaborated by the representatives of Neodarwinist evolutionary biology. The book is definitely worth reading.
20. Mendel's reserved attitude toward Darwinism is discussed, e.g., in the *Proceedings of the Gregor Mendel Colloquium*, 161–172, 1971 and *J. Hered.*, 87: 205–213, 1996.
21. After completing this manuscript, I read it through several times and replaced all dispensable professional terms and foreign words by their general equivalents. Sometimes, this was very hard for me, as I was aware that, in making the text easier to read, I was reducing its exactness. In the book, I have attempted to leave in only the absolutely necessary professional terms, which the non-biologist reader will, unfortunately, just have to get used to. For example, the terms "become fixed, allele, genetic variability, gene, genotype, genome, gene pool, heterozygote, homozygote, phenotype" have been retained.

CHAPTER 3 How Darwinism became normal science and what the "New Synthesis" was

From the standpoint of Darwinism, the end of the 19th century was a period of stagnation, especially compared to the substantial progress made in the other biological and non-biological sciences. Darwin's originally clear conception of biological evolution requiring a central role for natural selection became increasingly misty and a number of authors even felt the need to noisily distance themselves from it.[1] At that time, few biologists doubted that biological evolution (the development of species from a common ancestor) actually occurs; however, the role of natural selection was thrown into considerable doubt. Natural selection was frequently considered to be only a sort of net that regularly removes individuals with detrimental deviations from the original phenotype but that cannot form any new structure or pattern of behaviour increasing the viability of organisms. Sources of adaptive traits were (unsuccessfully) sought elsewhere, for example in unknown forces driving evolutionary trends (see p. 191) or in the supposed ability of organisms to create adaptive mutations (mutations enabling an organism to meet the momentary requirements of the environment).

At the beginning of the 20th century, a great breakthrough came with the rediscovery of **Mendel's laws of genetics** controlling the transmission of genetic information.

In the following years, geneticists examined these laws in great detail and primarily found the reasons why they govern the transfer of genetic information.

Box 3.1 Mendel's laws of genetics

According to Mendel's first law (the **law of segregation**), in each generation, two alleles of any gene present in the parent individual segregate into independent sex cells (e.g. into individual sperm) without undergoing any change and without affecting one another. The second law (the **law of independent assortment of characters**) states that the individual pairs of alleles of various genes segregate into sex cells independently of one another and that the manner of segregation of one pair of alleles in no way affects the segregation of another pair. In the first decades of the 20th century, geneticists demonstrated that Mendel's second law applies only to pairs of genes, each of which belongs to a different chromosome (see also Box 4.10 on p. 59).

Why nature is governed by Mendel's laws, or a little secondary school material won't hurt you (I hope)

Gradually, everything that can now be found in secondary school textbooks was discovered. Thus, it was found that genes are mostly localized in the nuclei of cells, where they are arranged one after

Box 3.2 Chromosome

Genetic information is written in the DNA molecule in the cells (see Box 3.6). Human DNA in the cell nucleus has an overall length of about two metres. In order for it to fit into the cell, it is wound around specialized proteins (histones) and, together with them, folded many times and wound in chromosomes, rod-like shapes usually with a length of several thousandths of a millimetre. For example, human beings have 46 of these species, which differ in size and shape, in the cell nucleus. Each chromosome is formed of two identical **chromatids**, whose DNA was formed by copying the chain of a DNA molecule originally contained in one chromatid (Fig. 3.1). During nuclear division, the two chromatids separate to the opposite ends of the cell, ensuring fair (even) distribution of the genetic material between the two daughter cells. When cells are not dividing, the chromosomes are loosened and are not visible without using special microscope techniques. They change into their characteristic form observable under a normal (optical) microscope during cell division.

another in a number of pairs of rod-like bodies, called **chromosomes** (Fig. 3.1). Each of a pair of chromosomes contains the same set of genes, but they usually differ in the variants of these genes (their alleles). At a certain site, one chromosome can have an allele determining brown eyes, while the other has one for blue eyes. All the cells of the body of "respectable" multi-cell organisms (e.g. man) contain the same set of genes. Sex cells in sexually reproducing organisms are an exception; they are usually formed in special organs as the result of a special type of cell division. Only one

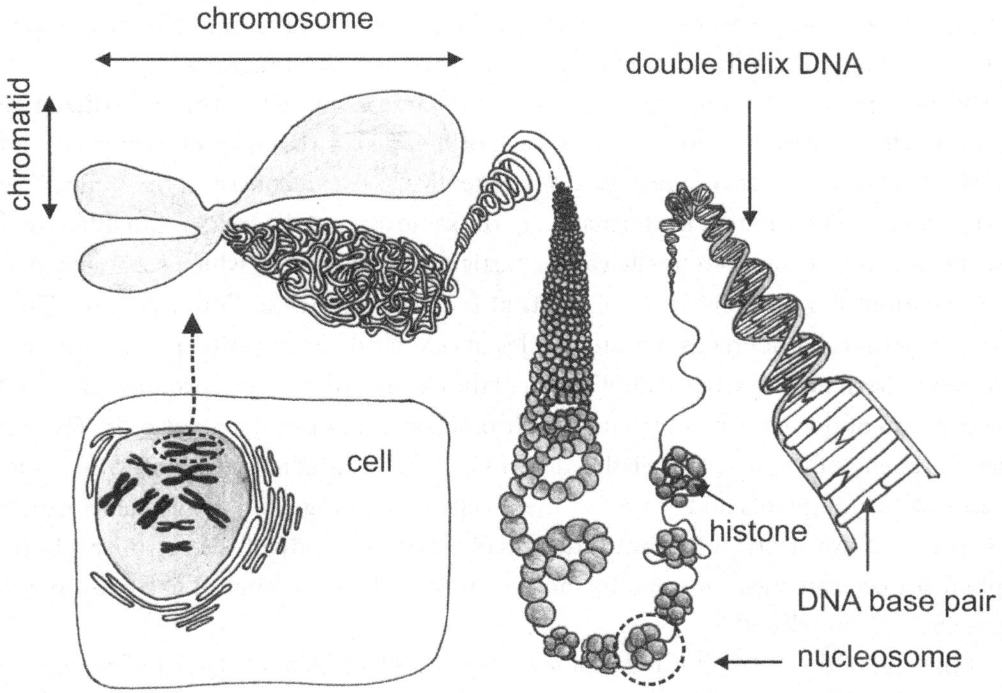

Fig. 3.1 Structure of the chromosome. Each of the two chromatids (connected at the narrowed site, i.e. the area of the centromere) is formed of a DNA strand wound repeatedly around nucleosomes consisting of specialized proteins – histones.

Box 3.3 Genome

Genome is the sum of all the genes occurring in the cells of a given individual. In contrast, genotype is the sum of all the alleles of a particular individual. The nuclear genome is the sum of all the genes occurring in the nucleus of the cell; the cytoplasmic or organelle genome is the sum of the genes contained in the DNA of cell organelles, mitochondria or plastids. The genomes of males and females of a certain species can differ in the presence or number of sex chromosomes, i.e. chromosomes whose occurrence or number determines whether the individual will develop as a male or female, see Box 8.6 on p. 16.

of each pair of chromosomes passes into the nucleus of the sex cell. Sex cells (in animal sperm and eggs) thus have a **haploid genome**, containing one copy of each chromosome and thus one copy of each gene. During the formation of sex cells in a particular individual, the genes of both his parents are randomly mixed as a consequence of two processes: genetic recombination and segregation. In **recombination**, first pairs of the corresponding chromosomes are formed (by fitting themselves together along their length) in the nucleus of the cell which, in the future, is to be responsible for the formation of the sex cells; then, within this pair, the chromosomes exchange some corresponding segments. In **segregation**, one chromosome from each pair is randomly allocated to the two newly forming cells. At this moment, each chromosome consists of two completely identical **chromatids** (see Box 3.2), so that it is only in further division, in which these chromatids separate, rather than whole chromosomes, that haploid sex cells are formed. Double division, through which haploid cells are thus formed, is termed **meiosis**.

During reproduction, the combination of two haploid sex cells forms a **fertilized germ cell** (zygote), which has a chromosome from each of them, i.e. two copies of each gene, and is thus a **diploid.** The two copies of the gene can be identical, i.e. can consist in the same alleles of the particular gene, determining the formation of the same form of the trait, or can differ. In the latter case, these can be two different alleles of a particular gene, each of which separately determines the formation of a different form of the trait (e.g. red and white flower colour). The diploid fertilized germ cell undergoes repeated cycles of cellular division (**mitosis**); to be more exact, it repeatedly alternates the stage of duplication of the chromatids in each chromosome and the stage of separation of one chromatid from each chromosome into two daughter cells. The number of cells – progeny of the original fertilized germ cell – thus gradually increases. In multicellular organisms (e.g. in plants and animals), these cells remain together, are mutually differentiated and specialize in various tasks, until they finally form the body of the organism. In this body, haploid sex cells are again formed by meiosis, with half the number of chromosomes, and the entire cycle is thus closed.

At the end of the 1920s, it was possible to derive both of Mendel's laws and explain a number of exceptions to their validity from the way in which organisms manage their genetic information (see Box 3.1 on p. 59). It is necessary to realize that Mendel derived his laws only on the basis of the numerical representation of individuals with different combinations of traits in the progeny

Box 3.4 Genetic recombination and segregation

These are processes that occur during the formation of sex cells. In these processes, a pair of similar, i.e. homologous, chromosomes in the nucleus forms doublets and mutually recombine. In **recombination**, the DNA molecule is broken at the same place in both homologous chromosomes. If the original parts of the same chromosome subsequently rejoin, no recombination occurs; however, if a strand of one chromosome joins together with a strand from the second chromosome, the pair of recombined chromosomes will differ in the combination of their alleles from the two original chromosomes.[2] While, prior to recombination, it was possible to state that one chromosome was derived from the father and one from the mother, the recombined chromosomes contain part of the alleles from the father and part from the mother. **Segregation** occurs during the separation of homologous chromosomes to the opposite ends of dividing cells. In this process, one of the chromosomes of each pair moves quite randomly to the opposite end of the cell. Even if recombination did not occur before this, the segregation of the chromosomes of paternal and maternal origin would give the newly formed cells their own combination of paternal and maternal alleles, different from the combination of alleles of either of its parents. Following separation of the pair of chromosomes in the first meiotic division, the two sister chromatids of each chromosome separate in the second meiotic division. Thus, four sex cells, haploid cells, can be formed from one diploid cell.

of two varieties of peas and this was thus only a **descriptive law.** Descriptive laws indicate how a certain phenomenon occurs (e.g. the numerical ratios in which individuals with various combinations of traits will be present in progeny), but they provide no information on why the particular phenomenon occurred in this way.

How Darwinism changed to Neodarwinism and how it became normal science

In the 1920s, **mutations** were discovered, i.e. changes in existing alleles to form new alleles, which occur very occasionally, for example, through the action of radioactive radiation or chemical mutagens. New genetic variability is formed through mutations and subsequent selection can then choose the variants useful for survival of the individual species. In the 1930s, this was followed by the discovery of the basic mechanisms of **speciation** – the formation of new species. At that time, biologists had collected all the stones for the mosaic, i.e. all the knowledge required for confirming the validity of Darwin's theory of evolution (however, they did not have, for the most part, the modern knowledge that could throw doubt on the validity of this theory or limit it). Nonetheless, until the 1930s, Darwin's theory tended to be rather a target of criticism amongst biologists. As was mentioned in the introduction to the chapter, almost no scientists doubted that biological evolution actually occurs, that species are formed one from another; but many doubted that blind natural selection could be the driving force for these changes and simultaneously the only cause of the formation of adaptive traits. It was not until the beginning of the 30's that the most influential evolutionary biologists concluded that, even if other mechanisms than selection are valid in evolution, natural selection is by far the most important and, basically, is in itself adequate for explanation of the development of species and the formation of adaptive traits. The combination of evolutionary biology and classical and population genetics (called the "New Synthesis") resulted in **Neodarwinism**.

Box 3.5 Stages in the development of evolutionary biology

Evolutionary theory based on Darwin's texts, elaborated at approximately the beginning of the 1920s, is generally called Darwinism or Classical Darwinism. The theory of evolution that was formed by incorporation of knowledge of genetics into the Classical Darwinist theory is termed Neodarwinism. Neodarwinists primarily understand evolution as a change in the representation of the individual alleles in the gene pool of the population and attempt to explain all evolutionary processes occurring at the level within and between species on the basis of this process. Consequently, for this reason, chapters devoted to population genetics – learning about the development of the genetic composition of the population – take up considerable space in textbooks of evolutionary biology. For most biologists, we are still living in the era of Neodarwinism. According to others, especially the work of S.J. Gould, who sharply differentiate between microevolutionary processes, occurring at the level of populations and species, and macroevolutionary processes, occurring above the level of species, and the gene-centred models of evolution following from the work of W.D. Hamilton, see Chapter 8, a new era in evolutionary biology has already begun. With my characteristic malice, I would like to introduce the term Postneodarwinism for this approach (and I look forward to seeing how my successors will manage to find a name for the next era of evolutionary biology).[3]

Basic textbooks were published in the field and any rejection or reduction of the role of natural selection in evolution began to be considered scientific heresy.

Without difficulties and, in fact, basically strengthened, Neodarwinism survived the important discoveries of molecular biology, including the discovery of the molecular structure of the gene, as an unbranched string of irregularly alternating groups of four nucleotides, A, G, C and T, in the double chain of the DNA molecule.

Traditionally, of course elsewhere than on the pages of basic textbooks in evolutionary biology, it was more or less possible to throw Darwinism into doubt in relation to the possibility of the occurrence of **targeted mutation**[4], or in relation to the aspect of heredity of acquired traits.[5]

Box 3.6 DNA

DNA molecules are fundamentally like two long strings of beads, each of which consists in irregular alternation of four types of beads – **nucleotides**, twisted in a helix, one around the other (Fig. 3.2). At the site where nucleotide A is present in one chain, nucleotide T is present in the second chain and where nucleotide G is present in one chain, nucleotide C is present in the other chain. If the two chains are separated, which can be achieved, e.g., by heating, the appropriate enzyme and all four nucleotides and a few other things are added, the appropriate complementary chain is formed according to the sequence in each chain so that, finally, instead of one DNA double chain, two identical DNA double chains are obtained (Fig. 3.3). This is essentially the basis for heredity of genetic information.

Scientific dissidents and other nonconformist souls from amongst evolutionary biologists were generously allowed to frolic to their hearts content within this playground. When they finally conclude that these phenomena seem to exist (and can occasionally substantially accelerate the adaptive response of the population or species to changes in the environment) but that the actual molecular apparatus that determines their functioning was created through the action of classical

Box 3.7 T.S. Kuhn

A historian of science (1922–1996) who was concerned with the laws governing development and scientific progress. He demonstrated that the concept of science as a regularly progressing process refining our knowledge is basically erroneous. He showed that three phases alternate in science. The phase of **normal science** is usually longest; here, slow development and refinement and elaboration of existing theories actually occur. This is followed by the phase of **crisis science**, when it is found that an ever increasing number of facts don't fit into the existing theory. The third phase is a **scientific revolution**, when the old theory is rejected and replaced by a new theory. The previous period and the previous theory are then forced out of the textbooks and subsequently from the consciousness of the relevant scientific community and, after some time, the history chapters of textbooks are rewritten as if the new theory had existed throughout all time. A new comfortable period of normal science begins. According to Kuhn, the main reason for this discontinuous development of science is the existence of **paradigms** – assumptions on which the accuracy of the theory stands or falls. However, during the period of normal science, scientists are not even aware of the existence of the paradigm and thus do not think about or test its validity.

Darwinist selection, they are frequently not only heard, but even treated with patience by the scientific public.[6] Evolutionary biology has become a **normal science** in the sense of the normal sciences of **Thomas Samuel Kuhn.**

Science, are you at all normal?

The scientists in a particular field share their **paradigm**, i.e. the central idea of the entire system of theories, ideas, the truth of which is not only not doubted and investigated, but which has gotten so far under the skin of them all that they are not even aware of it.[7] This allows them to collect knowledge and develop the field in an undisturbed, coordinated and effective manner. The shared paradigm allows them to recognize the important aspects at the given time, on which efforts should be exerted, agree on the truth or falseness of discovered solutions to these problems and agree on unimportant facts in the field and exceptions that need not be taken into account and which it is necessary to quickly and inconspicuously sweep under the carpet using **Occam's broom**. I think it was Sidney Brenner who first pointed out the existence of this convenient instrument, which is used in science almost as frequently as the much better known Occam's razor – though he obtained the Nobel Prize for his contribution to recognition of the laws governing the genetic regulation of the development of the organs and programmed cell death, and not for the discovery of Occam's broom, a hypothetical instrument which scientists use to sweep unpleasant facts under the carpet. I might go so far as to state that he obtained his Nobel Prize in spite of his discovery of Occam's broom. Scientists are not always pleased by a well-directed joke. However, if someone is really good and is willing to wait for recognition ten or fifteen years longer than would correspond to his results (and it would seem that such strange people do actually exist), then he can make jokes about his colleagues as much as he likes.

In the era of normal science, no acquired information can throw the validity of the central theory into doubt. Information that is contrary to this theory either ends safely under the carpet

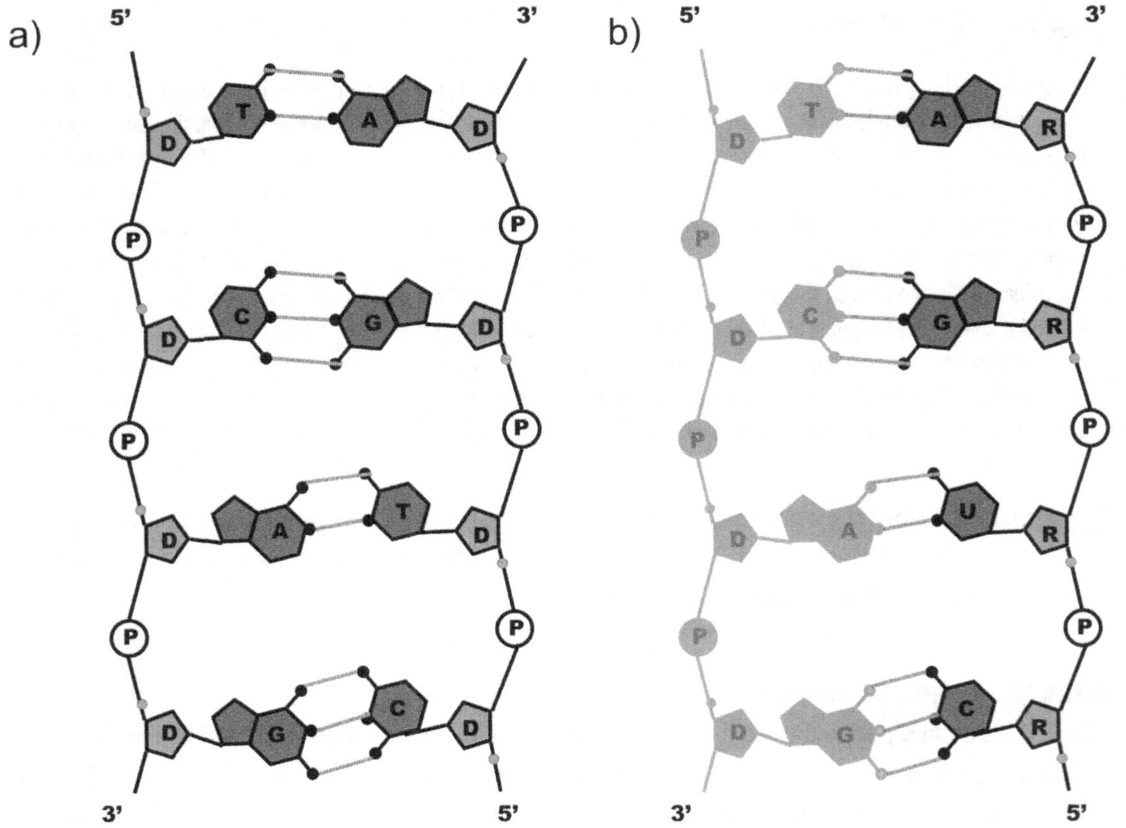

Fig. 3.2 The structure of DNA and RNA. The basic DNA unit (a) is deoxyribose sugar (D), which is bonded to one of four kinds of nitrogenous base, thymine (T), cytosine (C), adenine (A) and guanine (G). Two neighbouring sugars are connected by a phosphate bond (P). A sugar with a phosphate group and nitrogen base forms a nucleotide. The DNA molecule is thus formed of two linear chains in which four kinds of nucleotides irregularly alternate. The two chains are interconnected by hydrogen bridges between pairs of complementary bases, AT (two bridges) and GC (three bridges). The chains are mutually anti-parallel, i.e. the 5' end of one pairs with the 3' end of the other and forms a double helix structure in space. The structure of RNA (b) is quite similar to the structure of DNA. Ribose sugar takes the place of deoxyribose sugar and the pyrimidine base thymine is substituted in the relevant nucleotide by the base uracil. The figure depicts the RNA molecule at the moment of synthesis, when its base is paired with the complementary base of the relevant DNA section. However, during synthesis, the RNA molecule separates from the DNA molecules and is thus contained in the cell in its single-chain form.

or, if it is easier or if it doesn't fit under the carpet, leads to a slight modification of the theory – the theory is adjusted (mostly made more complicated) so as to encompass the new facts. A successful scientific theory can be recognized, amongst other things, because it is sufficiently malleable so that, if required, it can always be made more complicated to adjust to new knowledge. An insufficiently flexible theory disappears in the garbage heap of history, while more flexible theories remain.

In evolutionary biology, a number of facts have gradually been discovered that were not in accordance with the generally accepted theory and that were consequently more or less successfully swept under the carpet. As will be shown in subsequent chapters, some have disappeared from evolutionary biology entirely, while others have left a clear scar and others have even led to a certain modification of the existing theory of evolution. In this case, textbooks of evolutionary biology were extended to include an extra chapter. However, the chapter of the historical introduction and the overall internal conception of the fields have not undergone any changes. Basically, Neodarwinism in the form in which it crystallized by the end of the 1930s is understood as the last, final and only correct stage of the development of evolutionary learning. Theories that basically meant rejection of the correctness of the Neodarwinist model of evolution were presented to the public and generally accepted as a sort of cream on the cake. If fact, even their authors generally hastened to state that their discoveries are certainly not in any way contrary to the Neodarwinist model of evolution and do not reduce its validity.

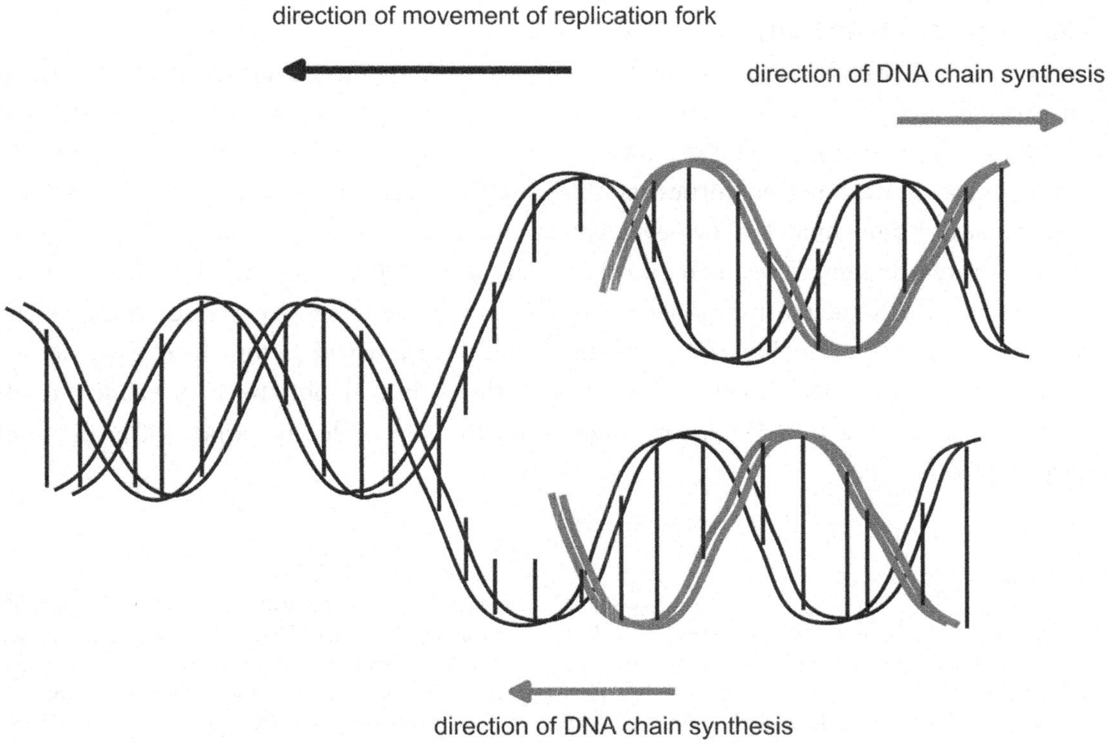

Fig. 3.3 DNA replication. During DNA replication, the two strands of the DNA double helix are locally unwound and separated using special enzymes and then each strand acts as a template for synthesis of the new DNA strand. Replication of the short genomes of viruses and bacteria frequently begins at one point, while the replication of long DNA molecules, encountered, e.g., for all eukaryotes, begins at many sites simultaneously.

Box 3.8 Theories and hypotheses

The destiny (or rather, the fate) of a theory is to be developed, i.e. altered over time in such a manner that it is gradually capable of encompassing and explaining more and more phenomena. In this, it differs fundamentally from a hypothesis. The destiny (or rather, the unavoidable fate) of most hypotheses, is to be **falsified**, i.e. to be rejected as invalid. Understandably, scientists would prefer to be able to **verify** their own hypotheses, to confirm their validity. I would like to emphasize the word "own" in the previous sentence. We very happily demonstrate the falseness of other peoples' hypotheses (and these are in the majority around us). Unfortunately, we must accept the unpleasant fact that scientific hypotheses (at least outside the field of mathematics) cannot be verified. For example, the hypothesis "all mammals give birth to live young" can be shown to be false if we encounter at least one mammal, for example, a duck-billed platypus, that hatches from an egg. However, if we did not discover a mammal hatched from an egg in books or nature, this would certainly not mean that we have confirmed our hypothesis. Until we study the reproduction of all mammals, extant and extinct, there still remains the possibility that such a mammal exists or existed (and that we have simply not found it) and that our hypothesis is thus invalid. As was convincingly explained by Karl Raimund Popper, hypotheses are thus divided into only two groups in science, the invalid, i.e. falsified, and the conditionally valid, i.e. those that have so far resisted attempts at falsification.[8]

Summary and incitement

It is perhaps rather unfortunate that we have not discussed anything fundamental that would be immediately related to evolutionary biology. We simply drew attention to the fact that the combination of Darwinism with genetics (i.e. the New Synthesis) led to the establishment of Neodarwinism in the nineteen thirties and this gradually became established as a normal science, in the sense of Kuhn's model of the development of science. The next chapter will be concerned with the above-mentioned cream on the cake, i.e. discoveries that, while they do not overturn the correctness of the Neodarwinist model of evolution, do to a certain degree reduce its importance for explaining the course of biological evolution. Specifically, we will discuss the role of chance in microevolution and macroevolution. Facts that throw into doubt the very validity of the Neodarwinist model, and that were consequently quietly swept under the carpet, will be discussed in later chapters.

Footnotes

1. If you want to read about how Darwinism is long dead, you could try reading the *History of Developmental Theory in Biology in the XIXth Century* (Dějiny vývojových theorií v biologii XIX. století) by Emanuel Rádl, J. Laichter (Prague), 1909, Engl. transl. 1932.
2. In actual fact, recombination occurs somewhat differently. At the beginning, for example, there is usually a single chain dissociation in one of the homologous chromosomes in the DNA. Subsequently, one end of the freed strand moves over to the second chromosome (in mitotic recombination, to the second chromatid), where it forces out one of the strands from the double stranded chain of DNA and replaces it. The replaced strand can shuffle back to the second of the pair of chromosomes and, with the assistance of reparation enzymes breaking down some section of the old strand and synthesizing new strands according to the corresponding section of the complementary strand, can transfer the particular topologically complicated structure to a large distance from the site of the original dissociation. The result of recombination can be not only that the alleles exchange places on the chromosomes,

but also that one allele "rewrites", according to its sequence, the alleles on the homologous chromosome – i.e. gene conversion.

3. Unfortunately, it was subsequently shown that I am not the first and, what is more, some of my predecessors took the introduction of the term postneodarwinism quite seriously. Well, it can't be helped, it is well known that good ideas arise at roughly the same time in different places (and also silly ideas, as can be seen).
4. Changes in the DNA that are not accidental in their effects on the properties of the organism, specifically, that assist the mutated individual in the situation in which it momentarily finds itself, to utilize the resources of the environment or to overcome its detrimental effects, are called targeted (or directed) mutations. According to the Darwinist theory, mutations are not targeted (i.e. have random directions), i.e. a mutation that is useful for the organism in the particular environment occurs with the same probability as a detrimental mutation. (In actual fact, most mutations tend to be slightly detrimental, followed by neutral and highly detrimental and only the smallest fraction of mutations are useful.) However, if we were to consider only a simple quantitative trait (such as body height) and include only mutations manifested in a small change in this height, roughly half the mutations would be useful and half would be detrimental under the given conditions. Even this is, understandably, a simplification because if, in the particular environment, large individuals were to have an advantage, not only mutations leading to a decrease in body height would fall in the category of detrimental mutations, but mutations leading to excessive increase in body height would also be detrimental (i.e. those whose consequences would "overshoot" the optimum height. Thus, there are certainly always more detrimental mutations than useful mutations.)

   A number of experiments performed in the past have shown that the vast majority of mutations are untargeted; i.e. they occur with the same probability under conditions where they are useful for their carriers as under conditions where they have no effect on their biological fitness. Of course, some mutations behave differently. For example, some organisms can apparently mutate (multiply) genes whose products are lacking in the cell (e.g. enzymes inhibited by a chemotherapeutical substance). If protozoa of the *Leishmania* genus are exposed to the action of methotrexate, a chemotherapeutical inhibiting the important enzyme dihydrofolate reductase, individuals rapidly appear in the culture that have a triply multiplied gene for this enzyme and are thus resistant to normal concentrations of methotrexate *Nucleic Acids Res.* 26: 3372–3378, 1998. Similar phenomena have been observed for mosquitoes of the *Culex* genus and even in mammal cells *Insect Mol. Biol.* 7: 295–300, 1998, *Cell* 37: 705–713, 1986. It is possible (and, I would say, quite probable) that the cells have formed a molecular instrument permitting targeted multiplication of genes whose expression (transcription to RNA) occurs with the greatest possible speed for long periods of time. The existence of such a molecular apparatus understandably allows organisms (more exactly, their cells) to undergo targeted mutation in many situations, as the fact that some genes are transcribed to the RNA with the maximum possible speed usually signal that the cell has a lack of its products and that its viability would improve if this gene were multiplied.

   Many phenomena that are or have been given in the literature as a manifestation of targeted (directed) mutation have in actual fact nothing to do with targeted mutation. For example, Cairns' mutations, i.e. mutations that occur over time in bacteria under conditions that prevent their reproduction, were interpreted as targeted mutations *Nature* 335: 142–145 1988, *Genetics* 128: 695–701, 1991. At the present time, most opinion is in favour of the explanation that this is a manifestation of the ability of bacteria to generate mutations even in non-growing cells (and apparently subsequently repair those mutations whose presence did not lead to renewal of growth) *Journal of Molecular Evolution* 40: 94–101, 1995. Bacteria that cannot momentarily divide because, e.g., an amino acid is lacking in the nutrient media, which they are not capable of synthesizing themselves, gradually mutate and again repair their DNA and do this before they exhaust all the resources and die out, or before they manage to "discover" a mutation that will renew their ability to synthesize the required amino acid. In the latter case, the mutated bacteria begin to reproduce and the useful mutation (which was originally on only one DNA chain) is not repaired.
5. The question of whether it is or is not possible to inherit acquired properties has disturbed and continues to disturb many evolutionary biologists. In organisms without a Weissmann barrier, i.e. where germ cells are formed from somatic cells, acquired properties can be inherited; however, it is not clear to what degree this ability can affect biological evolution. This aspect will be discussed in greater detail in Chapter 14. However, the possibility of inheriting acquired properties does not mean (as many people erroneously conclude) that Lamarckian evolution could function in nature. This assumes not only inheritance of acquired properties, but also that changes in behaviour cause changes in the

body structure of the organism that make this behaviour more effective. For example, that, when the predecessors of the giraffe reached up into the tops of trees, this would automatically lead to lengthening of their necks. Acquired properties can be inherited in plants (where there is no Weissmann barrier and where asexual reproduction is common), but they do not have a very wide repertoire of patterns of behaviour.

6. The following can be recommended for study: Jablonka, E., Lamb, M.J. *Epigenetic Inheritance and Evolution The Lamarckian Dimension* Oxford University Press (Oxford), 1995, *J. Theor. Biol.* 158: 245–268, 1992, Markoš, Anton: *Readers of the book of life: contextualizing developmental evolutionary biology.* Oxford University Press (Oxford), 2002, Riv. Biol.–Biol. Forum 94: 231–272, 2001.
7. In his famous book, *The Structure of Scientific Revolutions* (University of Chicago Press, (Chicago), 1996), Thomas Samuel Kuhn used the term paradigm with at least two meanings. This term can refer to a basic, very important idea, of whose existence the scientists in the relevant field can be aware. Thus, for example, it can be stated that a paradigm of modern molecular genetics consists in the transfer of genetic information from DNA through RNA to proteins. In its second meaning, paradigm seems to me to be a more useful term. It designates any necessary precondition for the validity of an important or unimportant hypothesis or model, of whose existence (and thus also potential invalidity) the authors (and the proponents and opponents) of the hypothesis are not at all aware. For example, the paradigm of the geocentric model of the solar system was that, if an object changes its position, it must move. For the proposal of the heliocentric model of the solar system, it was necessary to be aware that this paradigm may be erroneous, that if an object changes its position, it may just as easily be because we are moving relative to it.
8. Scientists are, understandably, aware that, in addition to hypotheses with a general quantifier, i.e. hypotheses expressed by the statement "For all X, it holds that Y." ("All mammals give birth to live young."), there are also hypotheses with an existential quantifier, expressed by the statement "There is at least one X for which it holds that Y." For example: There exists a mammal that is capable of obtaining all its energy by photosynthesis. We could verify this hypothesis if we discover such a mammal; however, we can never prove it false. It is obvious that we can never investigate all the mammals known to science and all those that science has not yet discovered. Then, is the basic statement of Popper's methodology about it not being possible to demonstrate the truth of scientific hypotheses erroneous? Of course not. In science, we are not concerned with the validity or lack of validity of the individual statements following from our hypotheses, but with the validity of the relevant hypothesis as a whole. The individual statements can have the character of statements with general and/or existential quantifiers, and thus can be proven either false or true. However, the statement determining the validity of the overall hypothesis is "All the consequences following from the validity of our hypothesis are valid." This is always a statement with a general quantifier and, as such, it is possible, in the best case, to prove it false, but it can never be proven true.

CHAPTER 4 Formation of differences between species – chance or necessity?

When Darwin sought an answer to the question of why species differ, specifically which process is responsible for the formation of **biodiversity**, he came to the conclusion that this is natural selection.

Box 4.1 Biodiversity

This term refers to biological diversity – heterogeneity. This has two components, on the one hand **diversity** in the narrow sense, which is the number of species and, on the other hand, **disparity**, which is the number of body plans of organisms and their difference. We can speak of local biodiversity, i.e. the diversity and disparity of species occurring in a particular territory or in a certain kind of habitat, and of global diversity, i.e. diversity and disparity of all the organisms on the Earth.

Although this famous book mentions the origin of species directly in its name, the aspect of the origin of species receives very little attention in it. It is frequently stated in Neodarwinist textbooks that Darwin did not consider this aspect at all. Neodarwinists frequently see the key point in the origin of new species in the creation of reproduction isolation barriers between new and old species, i.e. obstacles preventing the crossing of members of old and new species, rather that in subsequent diversification of species. Darwin, on the other hand, attempted primarily to find an answer to the question of why and how species diversify. Only a small part of the text is concerned with this aspect; nonetheless, it can be demonstrated that Darwin considered the resolution of this aspect to be of similar importance to the role of natural selection in the formation of adaptive traits. This follows both from his correspondence and also from the fact that the only figure in his book "On the Origin of the Species by means of Natural Selection" depicts the process of diversification of organisms through natural selection.

Darwin's (almost unknown) theory of the origin of the species

Darwin basically conceived that the species diversify as a result of the fact that very dissimilar forms developing in the framework of a particular species, i.e. extreme forms in the spectrum of intraspecies variability, are at a selection advantage compared to forms in the centre of this spectrum and thus, in great probability, these will become the basis for the formation of a new species. The middle forms are exposed to competition from both sides, while the properties of the extreme forms are so unique that they ensure a sort of monopoly on the use of part of the ecological niche of the particular species. If the mutually most different forms repeatedly lead to the formation of new species, the phenotype spectrum of organisms on the Earth will continuously increase and biodiversity (both in the number of species and their mutual dissimilarity) will increase over time.

The mechanism proposed by Darwin can perhaps function under certain, precisely defined conditions. In nature, we do actually occasionally encounter cases where selection provides an advantage for forms located at the edges of the phenotype spectrum of species and, on the other hand, places at a disadvantage to forms located in the centre of this spectrum. These situations occur especially when the particular species lives in a heterogeneous environment, where, e.g., two completely clearly defined life strategies can function well, i.e. requiring two different phenotypes, where the bearers of the transition phenotype are not at an advantage, even when they choose either of the strategies. A black butterfly can seem invisible on the dark bark of a spruce tree; a white form is well camouflaged on a white birch. However, grey forms will be visible on both spruces and birches, so will be caught by insect-eating birds.[1] This form of natural selection is called **disruptive selection**. However, **stabilizing selection**, i.e. situations in which natural selection prefers the middle forms and "punishes" the forms that deviate most from the usual phenotype, are encountered incomparably more frequently. Especially large and especially small individuals of a particular species usually have shorter lives and fewer progeny. This is quite understandable (in the past, evolution optimized the size of individuals in most species and a random deviation from the optimal value in either direction will tend to reduce the viability of the individual), but doesn't tell us much about the validity or invalidity of Darwin's hypothesis. As soon as a species is exposed to disruptive selection, it will probably sooner or later separate into several new species, each of which will be optimally adapted to a certain way of life and a certain environment. Individuals that differ from the relevant optimal phenotype in either direction are again at a disadvantage through natural selection. The number of species that submit to stabilizing natural selection should thus increase over time. However, nothing further of any significance occurs within species subject to stabilizing selection. If new species are to emerge somewhere, then this will probably occur within of a species that is exposed to disruptive selection for some reason. It is thus possible that, although stabilizing selection is encountered in nature far more frequently than disruptive selection, the latter plays a more important role in the evolution of biodiversity.

Although Neodarwinism tends to overemphasize the role of natural selection in evolution, Darwin's model of the divergence of species caused by natural selection is frequently treated rather superficially in modern textbooks. At the same time, it is apparently not possible to assess the validity of Darwin's model in the absence of detailed mathematical analysis. Neodarwinists were content to supplement Darwin's teaching with the **theory of speciation**, i.e. the aspect of formation of new species from old species. However, in this connection, they more frequently consider the aspect of the increase in the number of species rather than their mutual diversification. If the role of selection, or even disruptive selection, is discussed in this context, then this is almost always related to its role in ensuring reproductive isolation between the members of the newly formed species.

Box 4.2 Reproductive isolation

Most known species existing on the Earth reproduce sexually. Members of a single species cross almost exclusively amongst themselves. They mostly do not cross with the members of other species, or at least their crossing does not yield progeny. Barriers preventing crossing between species and thus ensuring reproductive isolation are basically of two types, external and internal. External barriers are formed, e.g., by mountain ranges, which separate the areas of occurrence of the two species; internal barriers consist, e.g., in the number and shape of chromosomes, which differ in the two species and thus prevent meiotic division, necessary for the formation of sex cells, from progressing to its conclusion. In some cases, it is not easy to decide which type of barrier is involved. For example, most biologists would classify as an external barrier the incompatibility caused in many species of insects by infection of part of the population by parasitic bacteria of the *Wolbachia* genus, which is capable of preventing reproduction of an infected individual with an uninfected individual or an individual infected by a different strain of this bacteria. However, if the cause of the infection were a virus hiding directly in the DNA of the cell, most biologists would probably consider the resulting reproduction barrier to be an internal barrier.

The part played by chance in making humans human-like (and apes ape-like)

Everyone who has leafed through a pictorial atlas of practically any group of organisms is probably aware that the mutual differences between the individual species occurring on the Earth are extremely obvious. Not everyone realizes simultaneously that the formation of this diversity of organisms constitutes a highly interesting and not entirely resolved aspect of evolutionary biology. A great many evolutionary biologists are now of the opinion that Darwin's explanation of diversification of the species through natural selection is not generally valid and that other processes play a more important role in this phenomenon. Discussions are continuing on this subject; nonetheless, it seems more and more probable that **chance** played an extremely important role in diversification of organisms and thus in the creation of biodiversity.

Macroevolution – the story of survival of the lucky ones

Chance certainly played a fundamental role in macroevolution and also substantially affects microevolution. **Macroevolution encompasses** all evolutionary processes occurring above the level of species, in the higher **taxons** (e.g. the rise and fall of the dinosaurs), while **microevolution** includes processes occurring within a species or its individual populations (e.g. spreading of strains of bacteria resistant to the action of antibiotics). New main branches of the **phylogenetic tree** are formed during macroevolution.

The members of this main branch differ in their **basic body plans**. The number of main branches on the phylogenetic tree and thus the number of basic body plans encountered in nature is not very large and was not very large in the past. Simultaneously, the number of conceivable and, from a functional standpoint, probably similarly good body plans is fundamentally infinite. Which of these possibilities actually occurred in phylogenesis is purely a matter of historical chance. Amongst other things, it follows from this that, on other planets on which life occurred, there will in all probability be completely different types of organisms than those that developed on earth.[2]

Box 4.3 Phylogenetic tree

The phylogenetic tree, or phylogram, is a graphical representation of **phylogenesis**, the gradual divergence of species from a common ancestor. In addition to the order of divergence of the individual species, a time scale can also be designated on the phylogenetic tree, permitting dating of the individual events in phylogenesis. For some purposes, it is useful to denote changes in the properties of the studied organisms on the phylogenetic tree, termed **anagenesis**. If the graph is used to depict not genealogical relationships between organisms, but their mutual similarity, then this is called a **phenogram**. Mutually unrelated species living in a similar environment and exposed to similar selection pressures (fish, dolphins, sharks, ichthyosaurus) can gradually become more similar (i.e. converge to a similar body structure) and can be placed close to one another on a phenogram (but not on a phylogram).

Another reason why chance played such an important role in macroevolution is the low number of species that had the opportunity to encounter one another on the Earth and to mutually interact, for example, compete for resources, during evolution. This is most evident when compared with the conditions under which microevolution occurred. Of course, the individual species occur in different numbers and are present in nature for varying periods of time. However, if a species such as whales is not involved and if man had no say in the matter, there are frequently millions to billions of individuals. The duration of existence of a species, from the instant of its formation to its extinction, is of the order of millions of years, while the period of survival of individuals is usually somewhere between weeks and several years or decades – for the moment we will ignore thousand-year old sequoias (*Sequoia*) or the mycelium of honey mushrooms as old as the forest itself; such organisms are certainly not typical and, in addition, for honey mushrooms, it is difficult to decide where the individual begins and ends. Thus, in microevolutionary proceses, the individuals within a species have enough time (enough "generations") to test all possible types of mutual ecological and evolutionary functions that their environment, phenotype and genotype, offers.

The situation is completely different for macroevolution. The number of species that occurred on the Earth over time is lower than the normal number of members of a single species. Although it may not be obvious at first glance, the average period of duration of species (for rapidly developing mammal species, say five million years) is comparable with the total time of existence of life on this planet. While this time is estimated at 3.5–3.9 billion years, the first 3 billion years of evolution were rather boring, at least from the viewpoint of us, multicellular organisms, i.e. animals, plants and fungi. Practically all the interesting events in phylogenesis occurred during the last 500–700 million years. The chain of species – ancestors – reaching back from any of the present-day mammals to the joint ancestor of all vertebrates (probably looking like an obese leach)[3] might have no more than 100–200 links. It is thus apparent that, during this small number of steps, evolution did not have much time to try out various pathways of phylogenesis and to test which of them is objectively best. At each of the cross-roads of phylogenesis, evolution simply randomly selected one of the possibilities and then, with only the substantial assistance of natural selection within the species, "attempted" to maintain the particular phylogenetic branch as long as possible. Thus the body plans and taxons that we encounter in extant and extinct organisms depend primarily on the sequence of several hundreds (and perhaps only tens) of random decisions that evolution made at key phylogenetic cross-roads.

Microevolution works on an entirely different basis. While each mutation is also a random experiment, because evolution carried out an enormous number of these "experiments" during the existence of the species, natural selection is able to select the best of each of them, i.e. those that actually contribute the most to the viability of the members of a particular species.

Imagine the work of a gardener as a parallel to the work of evolution. When a gardener trims a bush that branches out close to the ground into a thousand new branches dividing again and again, for example snowberry, he can choose any final shape and mould the bush into this shape in several seasons using garden shears. However, if a gardener wishes to trim a bush that sends out a maximum of two or three shoots on each branch or doesn't form any new shoots at all, for example elderberry, he will have to substantially tame his creative imagination and accept the fact that the bush will decide on its own final shape, from his viewpoint, at random. Microevolution, i.e. evolution within a species, has a character that is closer to the work of a lucky gardener who can decide in advance which shape he will train a bush to adopt. On the other hand, macroevolution is more similar to the work of his less lucky colleague who, if he isn't to go completely crazy, must accept the fact that he can only assist chance and nature. (Of course, this similarity is not perfect – in contrast to the gardener, evolution doesn't plan anything in advance and only removes things as a result of the pressure of natural selection.)

Occasional catastrophic events leading, at a certain moment, to **mass extinction** of part of the fauna and flora, are another important source of chance in macroevolution. These catastrophes were caused, for example, by the impact of a large meteorite or the core of a comet on the surface of the Earth or extremely strong volcanic activity that occurred over a large area at a certain time.

Box 4.4 Taxon

A taxon is a particular complete part of the phylogenetic tree (branch or monophyletic group or clade), which the relevant professional – taxonomist – defines and names. Thus, a taxon can be a single species, such as a chimpanzee, or perhaps the family of canine carnivores. At the present time, it is required that each taxon be monophyletic, i.e. that it include only a single species (common ancestor), whose ancestor was, itself, not a member of the particular taxon. A large number of experts (cladists) also require that the taxon include all the descendants of a particular common ancestor. Thus, cladists declared that a number of former taxons were invalid, including such ones as fish and reptiles. (It must be admitted that they had quite good reasons for this; however, it is probably better not to discuss this here.[4]) A taxonomist can define and name any branch of the phylogenetic tree; however, in actual fact, he defines only those taxons whose members differ substantially in some way from the members of other taxons.

Box 4.5 Chance

We will leave the question of whether chance exists objectively, or whether all events occur according to certain laws, to philosophers. However, subjectively, chance certainly exists. We consider that all events, whose occurrence does not follow from the properties of the system that is the subject of our interest and whose behaviour we wish to explain, are governed by chance. For example, the extinction of the dinosaurs at the end of the Mesosoic as a consequence of the impact of a cosmic body was a chance event from the standpoint of a biologist, as it was not possible to derive in any way that it would occur from the properties of the organisms that occurred at that time on the Earth and from the laws governing the development of living systems. Simultaneously, it is of no importance whether or not it was determined at the instant of formation of the solar system that this cosmic projectile would collide with the Earth and thus whether it was or was not possible to predict that this would occur from knowledge of the positions and movements of the bodies of the solar system (and its surroundings) and on the basis of the laws of physics.

The extinction of up to 90% of species (i.e. species whose fossils have been preserved, thus primarily marine fauna with hard shells) was usually caused by drastic changes in the weather and in the chemical composition of the atmosphere and hydrosphere, or substantial changes in sea level, that occurred as a consequence of the particular catastrophe.

Simultaneously, the extinction of organisms is usually a matter of chance; it could easily have affected the previously most successful species and entire phylogenetic branches of organisms. It is irrelevant how well the species or particular higher taxon was adapted to the conditions prevailing in the particular area prior to the catastrophe. The species that survive the catastrophe is determined, at the particular moment, by properties that occur in the given species purely by chance, for example, the ability to survive for a long time in an environment with low oxygen levels or in an environment from which most competitive species have disappeared. This is called **preadaptation**, i.e. properties or abilities that are very useful for survival of the organism under the given conditions but that evolved in the particular species in the past for completely different reasons and as a consequence of the action of completely different selection pressures. While the wings of penguins are excellent for swimming, birds' wings did not originally evolve as an adaptation for swimming, but as an adaptation for flying. Birds' feathers were probably not developed as an adaptation for flying, but rather as a means of thermoregulation. Similarly, the ability to survive in an environment with a lack of oxygen almost certainly did not develop in any species as an adaptation for survival after the impact of a giant meteorite, but rather as adaptation to life in mud. Darwin's theory is an excellent description of microevolutionary processes. In accordance with it, in competitive battle within a species, those individuals that are best adapted to their environment survive best. On the other hand, in macroevolution, adaptation to environmental conditions plays only a secondary role. In the long-term time scale, species and phylogenetic branches survive whose individuals are preadapted to future catastrophes, i.e. conditions that they probably never encountered. **The stronger survive in microevolution, while the lucky ones survive in macroevolution.**

Chance in microevolution – what is up is also down

And what is worse (for Darwin). The results of molecular genetics obtained in the last third of the 20th century indicated that, even in microevolution, chance plays a very important role in the diversification of the species. Studies performed on various organisms have demonstrated that the order of the nucleotides in the individual genes and the order of the amino acids in the individual proteins differ so substantially between various species and between various individuals in the population of a single species that it is almost impossible for natural selection to be responsible for their mutual diversification. If, in a given species, a certain new mutation, for example the replacement of one amino acid by a different amino acid at a certain site in a particular protein, is to completely predominate (i.e. **become fixed**), all the carriers of the original variant of the particular gene must die out. The rate of spreading of the new advantageous mutation is thus limited from above by the rate of multiplication and dying out of the members of the relevant population. This maximum rate must be divided up amongst all mutations spreading in the population at the same time. If only one mutation that increases the biological fitness of its bearers by one quarter, for example, increases by one quarter the number of young born and spreads in the population, then about 1 000 years would be required to replace the original variants of the particular gene. However, if another advantageous mutation spreads in the population, e.g. reducing to one half the risk that its carrier will be caught by a carnivore, then the spreading of the original advantageous mutation will be much slower and would take, for example, 2 000 years. I don't intend to find out or calculate how great the slowing down would actually be (fortunately, I am no longer writing textbooks of evolutionary biology). You can try calculating this yourself and, if you do this for populations with various types of size regulation, you could even produce a work published in a journal with impact factor.

Thus, advantageous mutations compete in spreading and obstruct one another, so that they retard the spreading of each further mutation in the population. If, for example, the number of amino acid replacements differentiating two species that branched off from a common ancestor three million years ago is several orders of magnitude larger than the number of mutations that

Box 4.6 Biological fitness

This was originally an umbrella term expressing the overall ability of an individual to produce (fertile) descendants in comparison with the other members of the population. Neodarwinists assign biological fitness to individual alleles. In this conception, this is a number that expresses how many fewer descendants on average are produced during its lifetime by the carrier of a particular allele compared to the carriers of the most successful alleles in the particular population. The biological fitness (w) can be used to calculate a **selection coefficient** (s) as $w = 1 - s$. The selection coefficient is thus the obverse of biological fitness as it expresses the degree to which the bearers of certain alleles are affected by natural selection. If the carriers of allele A on an average leave the greatest number of descendants, i.e. 10, while the carriers of allele B leave an average of 8 descendants, then the biological fitness of carriers of allele B is 0.8 and the corresponding selection coefficient is 0.2.

Box 4.7 Journal with impact factor

A journal with impact factor is a journal that was included in a certain bibliographical data base (Science Citation Index) some years ago on the basis of a combination of coincidences, and since then has been considered to be more prestigious than some possibly better non-impacted journals. Each year, the database operators calculate an **impact factor** for particular journals included in this database; this is the average number of references to one article (see Box 2.6) in journals included in the database within the first two years after the publication of the article. Half of the journals had an impact factor of less than 1 in 2004; however, there were about 10 journals with an impact factor of greater than 30. The higher the impact factor of the journal, the better the articles published in it are considered to be when evaluating the performance of a scientific worker or scientific institution. Simultaneously, a substantial number of evaluators (and evaluated persons) apparently do not realize that the order of the journals would be completely different if the impact factors were calculated from the number of references, not within two years, but within four or even ten or fifty years after the publication of the relevant articles and that it is frequently not possible to statistically demonstrate a connection between the impact factor of a journal and the number of references to the individual articles published in it. (This apparent paradox is caused by the fact that the differences in average citation are mainly caused by differences in the probability of the occurrence of a few highly cited articles; most articles that are published in any journal are cited to roughly the same degree. To be more specific, 20% of biochemical or molecular biological works are apparently not cited even once even five years after publication, approximately 75% of articles in the social sciences are not cited and 95% of articles in the humanities, where there is a tendency to write and refer to books, are never cited.[5] The main contribution of the existence of a database of impacted journals thus does not consist in its usefulness for evaluation of the quality and quantity of scientific work, but in the fact that it reduces to a certain degree the scope for establishing an increasing number of scientific journals and thus permits concentration of the sources of scientific information in the already existing journals. After a certain period of time, a new journal can be included in the database of impacted journals; but its articles must be sufficiently cited beforehand. And who would send his good article to an "unimpacted journal"?

could accumulate by selection over the same period of time, it is apparent that other factors than natural selection must have played a role in their spreading. At the present time, it is assumed that these factors consist particularly in genetic drift and evolutionary hitchhiking, also called genetic draft. The **theory of neutral evolution** is concerned with the development of organisms through the accumulation of selectively neutral traits as a consequence of the action of the above processes.

Box 4.8 Theory of neutral evolution

This theory is concerned with study of the evolution of selectively neutral traits, i.e., for example, a large part of changes in the DNA sequence. As, in some cases, up to six various triplets of nucleotides code the same amino acid, a change in the DNA need not have any effect on the amino acid sequence of the protein that is coded by this DNA. Thus, mutations that do not affect the sequence of proteins can be invisible for selection and thus their spreading and accumulation in the genome must occur through some other process than selection. Traditionally, primarily genetic drift is considered; however newer discussions consider genetic hitchhiking (which may be more significant). Neutral evolution may be responsible for the evolution of a greater number of traits than selection alone (however, this is not entirely certain[6]) and can thus substantially contribute to the diversification of species and possibly also to speciation (the splitting off of new species). However, the most interesting class of traits – adaptive traits – cannot be created by the processes of neutral evolution.

What do mice do in the park when they have nothing to do? –They drift

Genetic drift is basically a different term for **chance**. Imagine a population of 50 mice living in a city park. The DNA of some of these mice has a mutation that means that the tip of their tail is black. It is assumed that the colour of their tails in no way affects the viability of mice or their ability to reproduce. In the spring of a particular year, the particular mutation will be present to a level of exactly 50%, i.e. the mutation will be present in 50% of the copies of the particular chromosome occurring in the given population and the remaining 50% of the chromosomes will bear an unmutated variant of the gene, responsible for normal tail colour. What is the probability that the representation of the particular mutation will be exactly 50% in the population in the spring of the following year? Certainly less than that it will differ from 50%. The representation of the mutation in the population will most probably increase or decrease slightly. This will occur purely by chance, for example because wandering cats would catch a few more or a few less individuals whose DNA bears the particular mutation than individuals without this mutation. If the population were very large, e.g. a million individuals, then the random effects would cancel out and the deviation from 50% representation would be relatively small. However, if a small population of only 10 individuals were involved, then a random change in the representation of the particular mutation from one generation to the next would be very drastic. For illustration, you can try tossing a coin a thousand times (oh, alright, a hundred times is enough) and counting how many times it falls heads up. Then toss the coin only ten times and again count how many times it falls heads up. In all probability, the result of the first experiment will be much closer to the theoretical value of 50% than the result of the second experiment.

Let's return to our population of mice and let's say that the percentage of the mutation increased in one year from 50% to 53%. What will be the probability that this will remain unchanged in the following year? Yes, you're quite right. Again very low. Through the effect of chance, i.e. through genetic drift, this will again decrease or increase. And thus, from one season to the next, the presence of the mutation causing black tail tips will drift up and down until it one day reaches 0% or 100%. In the former case, **evolutionary fixation** of the original unmutated variant of the gene with disappearance of the new evolutionary form would occur and, in the second case, evolutionary fixation of the new form, i.e. the mutated variant. As soon as evolutionary fixation occurs, neither drift nor selection can further change the representation of the individual gene variants.

What is the probability that genetic drift will fix a newly occurring mutation? It depends on the size of the population. This probability equals 1% in our population of 50 mice. Each mouse bears two copies of each gene in its cells, one from its mother and one from its father; thus, overall, each gene and each chromosome occurs in our population in 100 items (the fact that each mouse has a large number of copies of each gene and each chromosome and thus each gene as it has many cells in its body is irrelevant here). If there is sufficient time for drift and the park is not converted, let's say, into a more lucrative shopping centre, sooner or later 99 of 100 of the copies of the gene originally present in the population will disappear and all the mice in the park will bear the descendants of a single gene in their DNA. Of course, it is not possible to decide beforehand which of the one hundred originally present copies of the gene this will be; however, we can be certain that this will happen sooner or later and we can calculate how long this will probably take – in an ideal population, where there is, e.g., the same number of males and females, this will be in an average of 200 generations. When a new mutation appears in the population, it appears only in a single gene, on a single chromosome in a single mouse. The probability that this mutated gene will be the chosen lucky one that is fixed in the population by drift is thus 1:99, i.e. the above-mentioned 1%.

At first glance, it might seem that genetic drift is a very weak and slow process and that it will not play an important role, especially in large populations. This is not the case. The number of newly formed mutations in all the genes of the organism and in all individuals in the population together is so large that, even if most of them are never fixed, the small percentage of them that are fixed by drift is sufficient to be greater than the number of mutations fixed by natural selection over the same period of time.[7]

The hitchhiker's guide to the micro-evolutionary galaxy

Evolutionary hitchhiking or **genetic draft** is another process that can lead to very effective and, in this case, very rapid fixing of genetic differences between populations and also species. Some authors are even of the opinion that most mutations that can be observed at the level of the DNA are fixed by draft and that this process could be the main factor responsible for mutual diversification of the species, if they are assessed only on the basis of the number of differences observable in the DNA molecules.[8] As mentioned elsewhere, genes do not wander freely around the cells but, put simply, are threaded like beads one after another on the string of DNA in the chromosome. Genes that are located close together on the chromosome have similar fates to a great degree. When a certain mutation is created in a particular gene, for example, increasing the fertility of its carrier by 10%, then, in time, selection will cause not only fixation of this mutated

Box 4.9 Genetic linkage

The strength of the genetic linkage measures the probability with which recombination will occur between two genes on a chromosome. This is determined by the distance between the location of the particular genes on the chromosome and also the frequency of recombination at the given site on the chromosome. The existence of a genetic linkage is the reason why the behaviour of many pairs of genes is not governed by Mendel's second law, i.e. the law of independent combinability of predispositions. The strength of a genetic linkage can be measured from the ratio of the number of descendants in which recombination occurred between the particular genes and the number of descendants without recombination in this section. If there is the same number of both types of individuals in the progeny (for example, if the genes are located on different chromosomes), the genetic bond is zero; however, if the genes are close together on the same chromosome or if recombination does not occur in the area between the genes for some reason, the bond between the genes can be practically absolute.

gene but also, with a certain probability, fixation of the alleles located nearby in the chromosome. The further apart the genes are in the chromosome, the looser is the connection of their fates, as they will most likely be separated by the process of recombination. To make it more convenient for the reader, I repeat that genetic recombination occurs during the formation of sex cells. During recombination, a pair of corresponding chromosomes originally derived from the father and from the mother fit together, at some sites their neighbouring DNA strands are broken and the relevant sections derived from the chromosomes of the mother and father exchange places and the strands of these recombined chromosomes are rejoined – see also Footnote 2 in Chapter 3. Thus, the order of the genes on the chromosomes remains the same, but the combination of alleles on the individual chromosomes changes. The further apart two genes are on a chromosome, the greater the probability that recombination will occur in the section between them during the formation of sex cells. On the other hand, the closer two genes are on a chromosome, the stronger is the **genetic linkage** between them and the greater the probability that the alleles of these genes will share their evolutionary fate.

If a gene with a new, advantageous mutation is located very close to genes with selection-neutral mutations (i.e. mutations that in no way affect the viability of the individual) or with slightly harmful mutations, then there is only a low probability that they will be separated by genetic recombination. Consequently, the fixation of an advantageous mutation is accompanied by the fixation of these other (neutral or slightly harmful) mutations. This is, understandably, true

Box 4.10 Why do elephants change faster than mice?

Compared to small rodents, elephants have a much longer generation period. Nonetheless, palaeontological data indicate that they changed much faster during evolution than, for example, mice. The theory of genetic draft could provide a possible explanation for this. Because of their longer generation period, elephants live in a sort of rapidly changing world. During one generation period of mice (two months) the environment (for example, also the climate) doesn't change much (this does not apply to cyclic changes related, e.g., to the seasons of the year, but rather to long-term changes to which species react in evolution); however, substantial changes can occur during the generation period of an elephant. Consequently, elephants must adjust to new conditions in each generation and consequently new suitable alleles are quite frequently fixed in their populations. And neutral and only slightly harmful mutations hitchhike along with them; these need not affect the appearance of the elephant but can increase the probability that they will evolve into a new species, see the hypothesis of the formation of reproduction barriers as a consequence of accumulation of incompatible mutations on p. 84. Other explanations are also possible. For example, large animals usually form small populations; accident plays a more important role (compared to selection) in small populations, so that slightly harmful changes can accumulate more easily and thus faster in these populations. Both these phenomena are employed to explain the **paradox of the molecular clock**. Although most mutations are formed in copying DNA and thus in cell division during the formation of sex cells, the speed of the protein molecular clock, i.e. the speed of accumulation of mutations in proteins, does not depend on the generation period of the studied organisms. This is a result of the fact that, although fewer (slightly harmful) mutations are formed each year in elephants than in mice, a greater percent of them are fixed in elephants. In accordance with these hypotheses, synonymous mutations behave differently and their accumulation is not proportional to time measured in years but to time measured in generation periods.

only for mutations that are sufficiently advantageous and are thus so rapidly fixed that the fixation time is too short for recombination to occur in the section between them and the other mutations in the adjacent part of the chromosome. Thus, fixation of an advantageous mutation by selection can, through the mechanism of evolutionary hitchhiking, cause the fixation of a number of other selection-neutral mutations that would otherwise be fixed only with very low probability through the slow process of genetic drift.

Evolutionary hitchhiking is not driven only by fixation of suitable mutations but also by removal of harmful mutations. If a neutral mutation is located close to a harmful mutation that is removed from the population by selection, the neutral mutation will most probably also be removed with it. While the disappearance of variability in the section of chromosomes neighbouring on an advantageous mutation is usually designated as **selective sweeping**, regular removal of disadvantageous mutations accompanied by removal of other mutations in their vicinity is called **background selection**. Both processes lead to a similar result in that they, in the final analysis, lead to a reduction in the genetic differences between individuals in a single population and simultaneously to an increase in the differences between the different populations and species. Selective sweeping is a fast and effective process, as the occurrence of a single positive mutation sweeps out the variability present at the given site of the relevant chromosome in all individuals in the population. In contrast, background selection affects the variability of only a single specific chromosome in a single specific (mutated) individual, which dies out or does not multiply (or multiplies less than the average multiplication of its competitors without the harmful mutation). However, there are far more harmful mutations than useful mutations. Mutations are random in the direction of their effect. A random intervention into a functioning system can improve the functioning of this system, but will tend rather to worsen it. For example, if you were to replace randomly selected words in the text of this page by other randomly selected words, this could improve the comprehensibility and maybe even the actual correctness of the final text, but there will be a much greater probability that it will be worsened. (At least I hope so.) Because of the great frequency of harmful mutations, background selection is at least as important as selective sweeping in removing internal species variability. From the standpoint of fixation of neutral mutations, understandably hitchhiking with positive mutations is the only important process.

And now, I will probably have to tell the truth. So far, I have managed to more or less successfully hide (squeeze into the box with the elephants) the existence of probably the largest category of mutations. This is the category of **slightly disadvantageous (harmful) mutations.** These include mutations that slightly worsen the biological fitness of their carrier; however, this worsening is so small that their fate is determined rather by genetic drift than selection. It can be shown that this class includes all mutations whose selection coefficient in absolute values is less than $1/N_e$, where N_e is the effective size of the population.[9]

It is probably better not to know what the **effective size of the population** actually means; it is sufficient to bear in mind that, if a given population had the same number of males and females, they all had a similar number of young, the populations were not to increase or decrease and they all crossed mutually quite at random, the effective size of the population would equal the number of its members. All these conditions are usually not met, so that the effective size of the population is usually smaller than the number of individuals in the population. Sometimes it is very drastically smaller – for example, the effective size of a population consisting of one male and a million females would, surprisingly, be approximately the same as that of a population consisting of two males and two females.[10]

In small populations, slightly disadvantageous mutations form a very significant fraction of all mutations, and actually behave like neutral mutations. For example, in the above population of one male and a million-strong harem of females, a mutation that would decrease or increase the biological fitness of its bearer by 25% would behave as selectively neutral. On the other hand, in large populations, a relatively high percentage of mutations cross the borderline above which their fate is determined by selection rather than genetic drift. As most of them are harmful for their carrier rather than useful, they have only a small chance that they would be fixed in the population by selection or drift. However, they have a correspondingly greater chance that they will be fixed by evolutionary hitchhiking. In large populations, there is a quite large chance that an advantageous mutation will appear occasionally, which would be fixed by selection. Neutral and slightly disadvantageous mutations, which would otherwise be fixed by genetic drift only with low probability, are fixed each time together with these advantageous mutations.

Summary and incitement

Darwin was of the opinion that the driving force for mutual diversification of the species lies in natural selection, specifically the form that is currently termed disruptive selection. Neodarwinists demonstrated that other processes play a much more important role from the standpoint of diversity. Primarily, accident is important in both microevolution and macroevolution. During macroevolution, accident leads both to the formation of a relatively small number of key evolutionary innovations and also to random extinction of species and entire developmental branches, which occasionally occurs as a consequence of natural catastrophes. In microevolution, this is manifested in both genetic drift, i.e. entirely random fixation of some alleles, and in genetic draft, i.e. fixation of neutral and sometimes even harmful alleles that happen to be located close to advantageous mutations on the chromosome. The next chapter will be concerned with newly discovered evolutionary processes that are thought at the present time to be potentially responsible for the formation of complexity and organization in organisms. We will demonstrate that complexity and organization are two completely different properties and that organization is far more typical for organisms than complexity. We will further show that these properties were not formed in evolution only through the effect of selection, as Darwin and most Neodarwinists apparently thought, but also through the action of other forces.

Footnotes

1. The example given of industrial melanism is almost a compulsory part of every textbook on evolutionary biology. It has actually been independently documented in a number of places that, in areas where light-coloured tree lichens have disappeared as a consequence of industrial pollution, dark (melanic) forms of the peppered moth *Biston betularia* have gradually come to predominate. For example, in Manchester, the dark-coloured form gradually increased in number from zero to 98% of the population between 1848 and 1895. When the relevant legislation was introduced to protect the air and industrial pollution was reduced, the situation returned to normal and the light-coloured form again predominated. The case of the increase and decrease in the degree of industrial melanism is instructive and documentation by coloured photographs of moths with outspread wings pinned to black and white bark is certainly very didactic. In actual fact, the situation in relation to industrial melanism is quite different. It has been found that moths do not usually rest during the day on the trunks of trees, but rather in the dense twigs of coniferous trees, and certainly don't rest with outspread wings. The predominance of the dark forms of moths in polluted areas is almost certainly connected with industrial exhalations, but the chain of immediate causes and effects will probably be more complex (and quite certainly different) than is usually described in textbooks. *Journal of Evolutionary Biology* 13: 155–159, 2000, *Journal of Heredity* 93: 86–90, 2002, *Trends in Ecology & Evolution* 18: 640–647, 2003.
2. Not all evolutionary biologists and palaeontologists share the opinion that chance plays a fundamental role in macroevolution and that, if we were to "rewind the tape" to the beginning (like if we were to destroy all the organisms on the Earth with the exception of bacteria), completely different types of organisms would develop on the Earth than those we know at the present time. The fact that the universe around us seems to be silent, that we have not yet managed to capture any radio signals indicating the existence of intelligent life, is probably an indicator of the fact that at least some forms of life appear rather rarely in the universe. (It is, of course, quite possible that they are formed quite frequently, but also disappear very rapidly. I hope that the first alternative is true.) The opposite opinion (suggesting a much smaller role of chance in macroevolution) is put forward in the works of S. Conway Morris, e.g. *Geobios* 32: 165–174, 1999.
3. We so far do not know what the common ancestor of all vertebrates looked like. For quite some time, we thought that it could have looked something like a lancelet (*Branchiostoma*), i.e. like something between a headless fish stuck in the sand and a leech. However, molecular phylogeneticists have recently shown that a lancelet is related to us less than, say, sea squirt. It is rather improbable that our direct ancestors would be similar to a sessile sea squirt. So we had better remain with a fat leech…
4. Cladists hold the principle that a valid taxon must be monophyletic, i.e. it must contain only species that are more related mutually than to any arbitrary species that does not belong in this taxon. Of three species, they consider that those two species that had a common ancestor, which was not simultaneously an ancestor of the third one, are related. Consequently, cladists cannot consider that the taxon of anthropoids is valid, because a chimpanzee and human being (which is not included amongst anthropoids) had a common ancestor, which was not a common ancestor, e.g., of gorillas. Not all biologists share the opinions of cladists. Some of them are of the opinion that the requirement of monophyly of taxons is quite justified, but that it is not related directly to the maximum relationship between the members of the taxon, but to the number of its species whose immediate ancestor was not a member of the particular taxon. A taxon is monophyletic and thus acceptable (i.e. created properly) if it includes only one such species; a taxon is polyphyletic (and thus incorrectly created) if it contains or in the past contained two or more such species. While cladists attempt to ensure that the system they create reflects, as well as possible, the distribution of new evolutionary features (i.e. synapomorphy) in the individual branches of the phylogenetic tree, noncladists (called evolutionary or eclectic taxonomists) "prefer a compromise" and are interested both in sharing of evolutionarily new traits and in sharing of evolutionarily old traits (synplesiomorphy).
5. See, for example, *Nature* 423: 479, 2003. Understandably, if we were to analyze a sufficiently large data set, we would certainly find a significant correlation between the impact factor of a journal and the average citation of its articles. However, it is not clear how much percentage of the overall variability in the average citation of articles would be explained by the effect of the impact factor (how large the coefficient of determination, R^2, would be).
6. It quite clearly follows from theoretical models that most mutations that are found in DNA could not be fixed by natural selection. Nonetheless, this aspect is still considered to be unresolved. When the selection coefficients corresponding to the individual changes are studied on real data (mostly on synonymous muta-

tions, i.e. mutations that do not appear at the level of the proteins because of the degeneracy of the genetic code), they are almost always different from zero. This is also true when we compare the probability that the individual synonymous or nonsynonymous mutations will be maintained in the population in the polymorphous state or that they will be fixed in various species. The results of such comparison frequently indicate that synonymous mutations are also a subject of selection. I fear that the problem lies with the empirical data (or, more exactly, their interpretation) rather than with the theoretical models. To begin with, it cannot be expected that synonymous mutations should be, on average, selectively neutral. In addition to information controlling the structure of synthesized proteins, DNA and RNA also contain information required for regulation of transcription of DNA to RNA and translation of RNA to proteins and also information determining the rate of decomposition of RNA. From this, it follows that a great many synonymous mutations will have a substantially greater effect on the phenotype of the individual than nonsynonymous mutations. Thus, if we average the selection coefficients of these important mutations with the other (selectively neutral) synonymous mutations, we find that selection occurs everywhere. If, in addition, we forget about the existence of evolutionary draft and molecular drive and consider genetic drift as the only alternative to selection, we quite easily come to the (erroneous) conclusion that the theory of neutral evolution has no basis.

7. In fact, the same numbers of selectively neutral mutations are fixed in small and large populations. In a ten-times as large population, there is a ten-fold smaller probability that a newly formed mutation will become fixed; simultaneously, however, ten times as many mutations are formed over the same time in this population (there are ten times as many individuals that can mutate). In actual fact, it is rather more complicated; in a small population, the fate of mutations is determined more by chance than selection, so that a large part of slightly detrimental mutations act here as selectively neutral. In a large population, the same number of selectively neutral, a lower number of detrimental and a higher number of useful mutations are fixed than in a small population. As more detrimental mutations are formed than useful mutations, overall more mutations are fixed in a small population than in a large population over the same period of time. Unfortunately, even this is a considerable simplification, as we did not consider fixation of mutations by genetic draft, which is more effective in large populations, see below.
8. J.H. Gillespie wrote about the aspect of evolutionary draft (and is also the author of the term evolutionary draft). For those who are interested in reading more on this subject, I can recommend the works in *Gene* 261: 11–18, 2000 and *Evolution* 55: 2161–2169, 2003.
9. This relationship was derived by Moto Kimura and is expressed by the formula $\mathrm{abs}(s) \leq 1/N_e$. It follows from this formula that both weakly positive mutations and weakly detrimental mutations behave like selectively neutral mutations in small and medium-large populations (their fate is determined by genetic drift). It also holds that the same mutation can act as selectively neutral in a small population and as selectively important in a large population.
10. This relationship between the number of males and females in the population and the effective size of the population is expressed by the formula:

$$N_e = \frac{(4N_m N_f)}{(N_m + N_f)},$$

where N_m is the number of males and N_f is the number of females in the population. Try substituting the value for a population formed of a million females and one male; those are certainly scrawls, huh?

CHAPTER 5 How are complexity and organization formed in organisms and what does this mean?

The **complexity** of living organisms is a very obvious property. According to Darwin, the complexity of modern organisms was gradually formed during evolution through the action of natural selection. Selection systematically favoured individuals that were better able to (more effectively) utilize the conditions in the environment and simultaneously were better capable of resisting its pitfalls. Consequently, organisms in which adaptive structures were formed accidentally and assisted them in survival, for example, organs, patterns on the surface of their bodies or patterns of behaviour, had a better chance of survival. At the beginning of evolution, these structures tended to be simple and fulfilled their function only imperfectly. However, over time, natural selection improved their ability to fulfil a particular function. As the functionality improved, there was frequently also an increase in the complexity of the individual organs, in their number, and in their diversity and thus the complexity of the entire organism. Thus, Darwinists consider an increase in the complexity of organs and organisms to be a more or less **necessary side effect of an increase in their functionality** during evolution. I don't intend to throw doubt on this attitude at this point (this will take place a few pages further on). Initially, it is sufficient to doubt the opinion that living organisms differ from nonliving systems in their greater complexity.

What is complexity? ...hmm, that is quite complex

The complexity of a system depends on the number and diversity of the elements from which the system is composed and also on the number and diversity of the relationships amongst them. The complexity of a system is reflected quite well in the smallest amount of information necessary for its description. However – the amount of information required to create a system may be smaller than the amount of information required to subsequently describe it. Anyone who looks at a small section of an image depicting the Mandelbrot set and compares its complexity with the simple equation required to generate the entire infinite complex image will understand that an enormous difference can exist between the information required to create and describe a certain image (Fig. 5.1).

In nature, we can frequently encounter systems whose complexity is substantially greater than the complexity of any living organism. For example, this could be a pile of dirt, Niagara Falls or the El Niño atmospheric phenomenon. Nonetheless, the complexity of these systems usually doesn't seem in any way shocking to us, i.e. provided we aren't required to model the processes in

Box 5.1 The Mandelbrot set

This is a set of elements that belongs to the plane of complex numbers (in the figure, the abscissa corresponds to the real part and the ordinate corresponds to the imaginary part of a particular complex number c) that, even after repeated substitution into the (recurrent) equation $z_{n+1} = z_n^2 + c$ (where $z_0 = 0$), does not exceed a value of 2. Some points in the plane of complex numbers exceed a value of 2 in the very first substitution into the equation, while this occurs for others only when the given procedure, i.e. addition to its square and substitution of the result into the right-hand side of the equation, is repeated many times. The number of these repetitions (iterations) required to exclude that a particular point belongs to the Mandelbrot set is depicted by the degree of grey in the figure. (A much nicer picture is created when the numbers of repetitions are depicted in various colours.)

this system, e.g. for long-term weather forecasts. The reason is simple. What we, in fact, admire in organisms is not their complexity but an entirely different property, basically quite the opposite. This is their **organization** (in the sense of their orderliness).[1] While a great deal of information is required to describe a highly complex system, to the contrary, only a small amount of information is required to describe a highly organized system. For example, if certain molecules of a substance are ordered in a quite regular crystal, the exact position of several molecules is sufficient for description of the entire crystal. However, if we were to describe the structure of a pile of dirt, incomparably more information will be required (Fig. 5.2).

What are the common features of living organisms and crystals? And why have whole generations of biologists tended to confuse organization with complexity? A highly organized and simultaneously highly complex organism and a highly organized and a relatively simple crystal are similar in what could be termed statistical improbability. The statistical improbability of the formation of a certain specific system, similar to its complexity, increases with the number of elements from which the particular system is composed. When we look at a crystal or organism, we are thrown into doubt as to whether they could be formed by the blind combination of random processes in nature. The same concept doesn't cause us any difficulties when related to a pile of dirt consisting of the same number of molecules as an organism or crystal. A pile of dirt can be formed in a million and one ways and, although the complexity of its internal structure makes it highly improbable that two identical piles could be formed in nature accidentally, there will be no substantial difference between individual piles. Consequently, it doesn't even occur to us to ask what miracle in the infinite number of possibilities led to the formation of just this single pile. In contrast, in organisms and, to a lesser degree, also in crystals, a similar question makes a certain amount of sense. The molecules from which an organism and crystal are formed could also be ordered in a million and one ways. However, these possibilities are not mutually equivalent. Only a minimum number of them lead to the formation of a viable organism or (as in a crystal) to the formation of a regular spatial units delimited by mutually parallel planes, e.g., right angles.

The statistical probability of the formation of a viable organism or crystal by the random clustering of molecules is extremely small. However, this low statistical probability provides no

Fig. 5.1 Graphical depiction of the Mandelbrot set (a). This set includes elements that belong to the plane of complex numbers (in the figure, the abscissa corresponds to the real part and the ordinate corresponds to the imaginary part of a particular complex number c) that, even after infinitely repeated substitution into the (recurrent) equation $z_{n+1} = z_n^2 + c$ (where $z_0=0$), does not exceed a value of 2. Some points in the plane exceed a value of 2 in the very first substitution into the equation, while this occurs for others only when the given procedure, i.e. addition to its square and substitution of the result into the right-hand side of the equation, is repeated many times. The number of these iterations necessary to exclude that a particular point belongs in the Mandelbrot set is shown in the figure by the darkness of grey. The areas in the plane of complex numbers for which 170 iteration steps still did not lead to exceeding a value of 2 and that could thus belong in the Mandelbrot set are depicted in black in the figure. This simple algorithm leads to an extremely complicated shape whose individual parts can be constantly zoomed in and further and further details can be found in them (b, c, d). The figures were created using the excellent program XAOS, which was available free in 2008 on the web site http://wmi.math.u-szeged.hu/xaos/doku.php.

information on the **actual probability** of the formation of the particular system. Atoms or molecules are ordered into the shape of the relevant crystal very easily, as this does not occur through random clustering, but as the consequence of physical forces following from their shapes and the charge distributions on their surfaces. The highly organized and apparently highly improbable structure of a crystal is thus, in actual fact, the most probable state in which the given molecules can exist. In the case of the highly improbable structure of an organism, biological evolution, of which the organism is a product, is the primary source of organization and statistical improbability. A specific organism, such as the dying cherry tree in my neglected garden, is the result of individual development (ontogenesis), during which an enormous number of molecules (usually initially organized in individual cells, but we won't complicate things with this here) in a predetermined time sequence adopted the relevant positions and thus created the final form of the organism.

Fig. 5.2 Complexity and orderliness. The paving in the left-hand part of the sidewalk in Viničná street displays greater orderliness, while the right-hand side is more complex. (In the spring of 2003, the city authorities substantially increased the orderliness of this system, at the expense of its complexity. It is probably now easier to walk on the sidewalk but, if it were not for irresponsible dog owners, it would probably be rather boring.)

Box 5.2 Cells

Cells are the basic living units of all contemporary organisms with the exception of viruses. In some species of organisms, the body consists of a single cell (single-cell organisms); in other species, i.e. multi-cellular organisms, the cells multiply repeatedly during ontogeny, are variously relocated and diversify until, compared with the microscopic cells, they form the enormous bodies of a multi-cellular organism. The human body is apparently formed of about 50 trillion (i.e. 5×10^{13}) cells.

The molecular apparatus controlling the course of ontogeny (the development of an individual progressing from a fertilized cell to an adult organism), i.e. the DNA in the chromosomes and the apparatus for "reading" the information contained in the DNA, is, however, a product of evolution, i.e. this was formed mainly through the action of natural selection over the long time that life has been developing. The statistical probability that a regular crystal or even a living organism, perhaps only a bacterium, would be formed at a certain instant by random clustering of the relevant molecules at a particular, specific point in space is so low that such an event would not occur even once in the whole time of existence of the universe. However, if certain physical chemical processes (electrostatic forces, Van der Waals interactions, etc.) participate in the creation of the same object, a highly organized crystal is formed within a few hours and evolutionary formation of bacterial cells with participation of natural selection occurs in a few hundred million years. The complexity of a crystal (the number and variety of its elements and the relationships amongst them) is undoubtedly much smaller than the complexity of bacteria. Thus, it seems to us (and, objectively, this is true) that the random formation of a highly organized crystal is much more probable than the random formation of highly organized bacteria. However, they both seem rather improbable. And this is why we so frequently and so willingly confuse complexity and organization. I repeat and emphasize: what has seemed to people from time immemorial to be so remarkable about organisms is not their complexity, but the statistically highly improbable organization of their bodies. Incidentally, once again, language has shown itself to be wiser than its users. It will probably not be accidental that we call organisms, organisms and not, for example, complexisms.

On the formation of complexity (and organization) through self-organization

Thus, organisms exhibit great complexity and high organization where, in spite of general conceptions, organization tends to be the more typical of the two properties. Where does this organization come from? Which processes are responsible for its formation?

When I was studying at the university in the 1970s, some of my teachers, who thought more deeply, rather surreptitiously told us a great secret (I certainly don't mean this as an insult to my teachers; teachers who did not think deeply, and these were unfortunately in the majority, only repeated the individual parts of other people's textbooks): "It is quite possible that Neodarwinists are completely wrong in explaining the formation of organization. Organization was not formed in living nature as a result of natural selection, but spontaneously as a result of the fact that, from the physical standpoint, organisms are open systems that are far from thermodynamic equilibrium." It wasn't advisable to ask what "far from equilibrium" means; biologists were mostly not capable of explaining this and physicists refused to discuss these matters with us in a language other than formulae. While, in closed systems, i.e. in systems that do not exchange mass or energy with their surroundings, organization can only decrease over time, in open systems far from equilibrium, i.e. in "**dissipation systems**", organization can, on the other hand, increase at the expense of the organization around these systems. And this is actually the case of organisms. "My dear students. Energy derived mainly from the Sun constantly passes through the biosphere and the individual organisms. In organisms and the biosphere, this energy is converted from light energy to its lowest form, to thermal energy and thus the organization of living organisms is maintained at the expense of organization of the solar system. When organisms die, cease to ingest food, energy ceases to pass through them, and their organization gradually decreases. Dust thou art, and unto dust shalt thou return. And, if this seems too abstract, try putting a pan of clean water on a hot burner and watching how, at a certain temperature, before the water comes to a full boil, a regular, highly ordered system of honeycomb shapes are formed spontaneously on the surface. This structure is organized quite spontaneously because, at a certain level of energy input, it best allows for its dissipation, i.e. transformation from a higher form to a degraded form of thermal motion of molecules. And if this is not enough, if you have a friend who is a chemist, ask him to mix a suitable solution of citric acid, potassium bromide, sulphuric acid and cerium ions and allow the Belousov-Zhabotinsky reaction to occur in a flat dish. The marvellous mobile coloured shapes that will be formed and disappear in the dish until the relevant components of the mixture are used up will be pure beauty. In fact, the 1974 edition of Scientific American, No. 230, illustrates this beautifully. Please, left circulate."

Not that our deep-thinking teachers and Ilya Prigogine, discoverer of the importance of dissipation structures would be wrong.[2] Everything they told us about dissipation systems was correct; however, it was only marginally related to the organization of living systems. The organization formed during the development of the body of an individual is certainly not formed by simple self-organization of the system to a state permitting the greatest possible dissipation of energy.[3] The structure of the body of an organism is determined primarily by the information encoded in the DNA of the germ cell (zygote) and, of course, also the other protein and nonprotein components of the nucleus, cytoplasm and membranes of the cell, mediating in the transfer of information encoded in the DNA to the structure of the body of the adult organism. The body is not only formed, but also maintained for a long time in its functioning condition by a wide variety of molecular and cellular "devices", from the individual enzymes repairing damaged DNA or membrane to entire organ systems maintaining the integrity of the organism and eliminating, for example, newly forming cancer cells. The way these "devices" work was determined in the past by evolution (especially natural selection) and not by the blind and invariable laws of non-equilibrium thermodynamics. Organisms can, of course, not act contrary to these laws. As soon as energy ceases to flow through an organism and the system becomes closed from a physical point of view, the body necessarily undergoes decay.

The laws of non-equilibrium thermodynamics are not decisive even for the creation and maintenance of the organization of the biosphere as a whole. Here, also, organization was created as a consequence of quite different processes and forces, specifically biological evolution, and is maintained as a result of the information contained in the organisms from which it is composed. Once again, it of course holds that, if the Sun, as the main source of energy on which the vast majority of terrestrial ecosystems are dependent, were to go out, the organization of the biosphere would rapidly disappear. (Yes, I have heard something about ecosystems in the vicinity of submarine volcanoes and black smokers that are not dependent on solar energy but on the input of chemical energy from the centre of the Earth and the hot biosphere that Thomas Gold thinks could exist in the depths of the Earth's crust – in the interests of maintaining the continuity and comprehensibility of the text, I am prepared to conceal even far more important facts from the reader.[4]) It is obvious that if organisms disappear, then their organization also disappears. The force that is required to maintain certain things (here, the organization and complexity of life) need not be the same force that caused the formation of these things. If heating the Earth to 1 000 °C would cause the disappearance of known forms of life, this does not mean that life on Earth was formed by cooling its surface from 1 000 °C to the present temperature. **The theory of non-equilibrium thermodynamics allows us to understand why the formation of organisms is not contrary to physical laws; however, processes other than those active in the self-organization of dissipation systems are responsible for the formation of organization and complexity during evolution.**

Snow and games – sorting from the standpoint of stability

On the other hand, **sorting from the standpoint of stability** can be an important process responsible for the formation of organization and complexity during evolution. If objects mutually differing in their stability, i.e. objects that rapidly disappear after their formation, and also objects with substantially longer lifetimes are regularly formed in nature, then quite naturally the number of objects with longer lifetimes will increase in the environment over time. The ratio of newly formed stable and unstable objects will not change over time; however, objects with longer lifetimes will survive into the present from the past. **Sorting differs from selection in that it does not require that the newly formed objects inherit properties from their predecessors.** In fact, new systems can even form quite independently of one another and there need be no descendant-predecessor relationship amongst them. Snow flakes could be an example. Flakes of every possible size and shape fall to the ground. Let us assume that the fractions of the individual types of snow flakes do not change with time. As some of them are more stable than others, as time passes, the fraction of the individual types in the snow flakes will change and there will be an increasing predominance of those that will be more stable at the given temperature and humidity. Unstable snow flakes will disappear and more stable ones will, in contrast, accumulate.

Sorting from the standpoint of stability is active, together with selection in the evolution of living organisms. Many of the properties that we encounter in modern plants and animals and that we naturally assume were formed by Darwinist natural selection could, in fact, be a result of sorting from the standpoint of stability. However, sorting from the standpoint of stability plays an even greater role in some artificial systems that are intended to model biological evolution. These systems include the Life Game, thought up by John Horton Conway, and also the NK model of logical networks, described in his work by Stuart Kauffman.

In the **Life model**, space is conceived in the form of a large chess board. As in any other rectangular chess board, each square neighbours on the other squares (cells) at its sides and corners. Each cell can be in two states, either black (live) or white (dead). Development of the system occurs in individual cells according to the following rules: as soon as a dead cell is next to three live cells, it comes to life in the next step. If a live cell is next to less than two or more than three live cells, it dies in the next step as if by loneliness or overcrowding. The state of a live cell next to two or three live cells does not change in the next step. On the basis of these simple rules, a system develops from the originally disordered state of randomly distributed black (live) and white (dead) cells to a much more ordered state. At various places on the chessboard, relatively large black shapes are formed, which grow or move about (Fig. 5.3). Some shapes form and, on various sides, regularly emit other forms, where the meeting of two forms on the surface can lead to the disappearance of both or only one of them, or several new separate forms are created after the collision. Thus, in the system, sorting occurs from the standpoint of stability, as a consequence of which quite specific, stable structures are formed, whose remarkable properties have been described in the extensive "lifelogic" literature (or in Wikipedia).

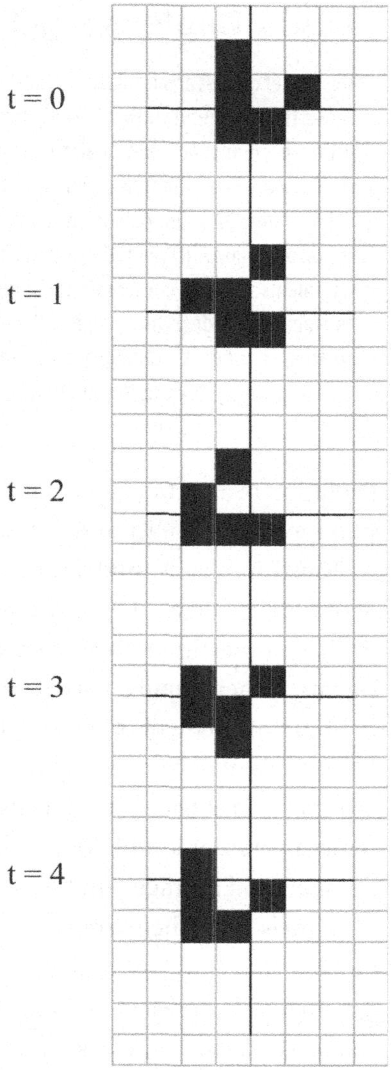

Fig. 5.3 Stable shapes are formed in the mathematical Game of Life. Amongst the patterns formed in the Game of Life, various types of "gliders" are quite remarkable, i.e. stable structures that move across the playing board. A "spaceship" remains stable until it collides with some other shape and until it reaches the edge of the playing board, i.e. it is cyclically converted into several shapes. The figure shows a simple glider that, in four steps, always moves one square diagonally downwards to the left.

Box 5.3 Boolean algebra

This is a set of rules and formal procedures using which we can derive the truth or untruth of a complicated statement consisting of a number of individual (true or untrue) statements. The individual statements are connected together by the logical operators AND, OR, XOR or the derived operator NOT. The logical operation AND yields the output TRUE if both the input statements are true. For example, the statement "It is snowing and raining outside" is true only if both individual statements are true, i.e. when it is raining and also snowing outside. The logical operation OR yields the output TRUE if at least one of the individual statements is true. For example, the statement "It is snowing or raining outside" is true if at least one of the individual statements is true, i.e. if it is raining outside or if it is snowing outside or if it is raining and snowing outside. The statement XOR is true only if one of the individual statements is true, i.e. it is raining outside or it is snowing outside, but it would not be true if it were both snowing and raining. If an (individual or complex) statement is introduced by the operator NOT, its validity is negated (a true statement becomes untrue and vice versa).

Evolution of a particular system is not Darwinian evolution, as its driving force is sorting from the standpoint of stability and not natural selection.

Kauffman's NK models are based on abstract, randomly created Boolean networks consisting of individual elements capable of transition between two states, on and off (true and untrue).

The properties of these elements, i.e. the manner in which they respond to a combination of signals at their inputs, correspond to the individual logical functions of a statement's logic. For example, an element of the AND type is converted to the "on" state only if activation signal "turn on" is present at both its inputs, and an element of the OR type if the activation signal "turn on" is present on at least one of its inputs, and an element of the XOR type if the "turn on" activation signal is present at only one of its two inputs. The individual networks differ in the number of elements and average number of bonds that connect the elements together, i.e. which transfer on-off signals from the outputs of one element to the inputs of another element. If an element is in the "on" state, the "on" signal is present at all its outputs; when it is in the "off" state, the signal "do not turn on" is present at all its outputs. At the beginning, one of the logical functions (e.g. OR, AND, XOR, etc.) is randomly assigned to each element and its state is random, i.e. on or off. The system again gradually develops in individual subsequent steps and, once again, complicated stable or unstable structures are formed in it. The unstable structures rapidly disappear and the stable ones accumulate in the system.

Although the Life Game and the Kauffman systems are often considered to be models of evolution driven by natural selection, in actual fact, the evolutionary processes occurring in these systems are driven by sorting on the basis of stability. While some structures formed in the Life Game are formed regularly and send out further shapes into the surroundings, these daughter shapes do not inherit the properties of the parent shapes. However, without heredity of properties, Darwinian natural selection cannot occur. The role of sorting from the standpoint of stability in biological evolution has not been extensively studied. However, it can be expected that this process could play an important role in the development of complexity and organization in living systems. It holds with high probability that simpler systems are formed more easily and thus more frequently than more complex systems. However, it simultaneously tends to hold that some more complex systems are more stable than simpler systems. The complexity of a system is, of course, no guarantee of its stability – we are regularly reminded of this, for example, by the manufacturers of can openers or creators of computer programs. (Oh, alright, Bill, XP is more stable than 98 (or at least it pretends this most of the time), but you must admit that they are both much worse than the ancient and once-cursed MS-DOS). **However, more complex systems can exist in far more variants than simple systems. And thus they can also include very labile, as well as very stable, systems.** The fact that labile systems disappear faster leads to a gradual increase in the content of more stable and thus more complex systems. Consequently, one of the side products of sorting on the basis of stability is also a gradual increase in average complexity and probably also organization of living systems.

On the formation of complexity (and organization) through passive evolution or tinkering in evolution

Another important source of organization and complexity in organisms lies in a process that could probably best be called **passive evolution**.[5] This is basically "patching up" mistakes occurring in biological systems during evolution. Biological evolution is not only a great improviser, but also a great "tinkerer". The elements of evolutionary tinkering are encountered constantly and their existence is one of the strongest arguments of evolutionists in the conflict with creationists.

What is involved? For example, the eyes of vertebrates. If we look at a secondary school scheme of the human eye, we are aware of something very special. It is, in a certain sense, turned around. Instead of the layer of light-sensitive cells facing frontward towards the source of light, it is directed backward towards the brain and the nerve cells that transmit the sight signals from these light-sensitive cells to the brain are, to the contrary, placed in front of the retina, in the direction towards the lens and the external environment. On their pathway to the retina, the light rays must find their way through a layer of nerve fibres, which certainly does not improve the sensitivity of the eye and the precision of the image. In addition, the bundle of nerves transmitting the signal from the retina cells must somehow get to the brain. Thus, there is a hole in the retina at a certain point, through which the bundle of nerves, a sort of biological "cable system", leaves the chamber of the eye in the direction towards the brain. Consequently, we have a blind spot on the retina, a relatively

Box 5.4 Creationist

A creationist is a person who, for some strange reason, believes that "belief" in the validity of the theory of evolution is incompatible with religious belief. Creationists consider that the best way to come to terms with this problem is to stubbornly convince the public that the theory of evolution is not valid. He would like to engage the public and politicians to exclude evolutionary biology from schools and, in the next step, probably from scientific institutions.

large place, where there are no light-sensitive cells and where optical signals can thus not be converted to the respective visual perceptions. If the image of a red stop light or a warning light on the dashboard is projected onto this place, this could have interesting and rather unpleasant results.

The fact that, in everyday life, the presence of a blind spot and the tangle of nerve fibers in front of the retina don't much bother us, is a result both of the fact that we usually have two eyes whose fields of vision overlap to a considerable degree, at least for us primates and owls (for that the image of the stop light that falls on the blind spot in the left eye falls on a functional part of the retina in the right eye) and also of the fact that the eyes move slightly all the time, so that various parts of our field of vision are projected onto the blind spot, and primarily because we have an extremely capable brain that is able to "calculate" a relatively good image of our surroundings from very poor signals provided by our less perfect optical sensors. In the absence of this excellent supercomputer and without the servomechanism of our eyes constantly twitching back and forth, our eyes would be practically useless.

If, in a store, they were to sell us a camera or video camera with similar quality optics, we would probably take it back after the first few pictures. And if they tried to tell us in the store that good-quality optics are not important, that it is sufficient to purchase a sufficiently powerful supercomputer to go with the camera, that would compute a perfect picture on the basis of poor input data, we would probably look at the salesperson in disbelief with our imperfect chamber eyes. And, it was basically this strange design that was selected by biological evolution for the eyes of vertebrates.

The reason is relatively simple. The predecessor of vertebrates, in which the eye developed, had a retina formed of a small number of light-sensitive cells, which almost certainly did not form a continuous layer. From the functional point of view, it made no difference whether the retina was connected to the nerves from the front or from the back. Evolution could "toss a coin" and choose one of these designs at random. In actual fact, chance apparently did not decide on the choice of one of the alternatives, but rather the laws of embryonic development of the eye – an eye with nerve fibers in front of the retina is easier to "produce" in vertebrates, where the nervous system, including the retina, is formed by shredding of a nerve tube. The impractical aspects of the originally selected structural design became apparent at a much later stage in evolution when, as a result of natural selection, the number and density of light-sensitive cells in the retina increased. Then the "cable system" began to reduce the quality of sight and enforced the formation of the blind spot. Evolution is not a designer who could sit down to a draft board and design a new, better design, such as an eye with the retina innerved from the back, i.e. the design that was successfully chosen by cephalopods. Rather, it set out to improve a basically unsuitable design, in this case through improvement of the organ intended for analysis of the optical signals obtained – the forebrain. The possibility that this evolutionary tinkering finally led to the formation of an extremely efficient brain that, as later emerged, can also be used in other ways than simply for

Box 5.5 The intelligent designer

This is a name for God that is used by a large portion of contemporary American creationists. They are concerned to circumvent the relevant article of the American Constitution prohibiting declaration of religious faith at state schools.

remedying the optical defects of an imperfect eye (for example, for conquering the biosphere of our planet, including the poor cephalopods with their structurally much better chamber eyes) is another story. We spoke about something similar in Chapter 3, when we were discussing preadaptations.

From the standpoint of discussions with creationists, it is important that evolution very frequently acted like a very stupid "intelligent designer".

In a great many cases, it short-sightedly chose the simplest but, in relation to future developments, a completely unsuitable technical solution. Evolution is a short-sighted opportunist that always resolves a momentary problem and never plans ahead and does not predict what negative consequences a particular constructional design will entail in the future. When the inadequacies of the chosen design appear, evolution "looks" (by the method of trial and error) for a way of minimizing their impact on the functioning of the system through various "patches" rather than looking for how to eliminate them through basic changes in the design. If we were to assume that organisms were formed not by blindly floundering evolution, but by an intelligent being, either God or, perhaps, an extraterrestrial, we would probably often be forced to doubt the level of his intelligence.

From the standpoint of the formation of organization and complexity, however, it is important that the **complexity of systems in evolution very frequently increases because evolution places one patch on another to remedy the individual mistakes rather than simply removing these errors**.

Thus, we can conclude that, because of passive evolution, very complex structures are formed in organisms, which are often objectively detrimental for the organisms. If they had not been formed in evolution, the particular species would be better off, i.e. it would be able to achieve greater population density under the same conditions. However, as we pointed out on the example of the eyes of vertebrates, in some cases a certain disadvantage (e.g., an imperfect eye requiring the formation of an extremely complex brain for its functioning) can, in time, unexpectedly become a basic advantage.

How (and why) should we edit RNA – let's ask trypanosome

A typical example of an extremely complicated system probably formed by this kind of passive evolution, i.e. gradually increasing layers of patches concerned to preserve the bare functioning of the system, is apparent in the molecular apparatus for **RNA editing**.[6] This process in its purest form is encountered in a great many members of protozoa of the Kinetoplastida order, which

Box 5.6 Organelles

Organelles are actually the organs of cells. These include, e.g., flagella or undulating membranes employed by the cell in locomotion, various sacs used to digest and transport food, the cell nucleus, in which the DNA is stored, and also **mitochondria and chloroplasts**, semi-autonomous organelles, residues of the originally symbiotic bacteria, producing energy in modern cells in the process of cellular respiration (mitochondria) or photosynthesis (chloroplasts).[7]

includes, for example, the etiological agents of sleeping sickness, *Trypanosoma.* The cells of these protozoa contain a kinetoplast, which is specially ordered DNA of an altered mitochondria. Mitochondria, which occur in the cells of all animals and plants, are the descendants of symbiotic (and perhaps parasitic) bacteria that the predecessors of modern cells originally "tamed" and learned to use to obtain energy from organic substances.

Similar to the mitochondria of other organisms, the altered mitochondria of trypanosoma contain their own DNA, the residues of the genome of the original bacterial cells. However, detailed study of this DNA has shown that it does not contain genes capable of coding functional proteins; more precisely, it contains these genes in a somewhat shorter, strangely ciphered form. Further research finally demonstrated that the molecular apparatus of trypanosoma is capable of deciphering these ciphered genes prior to use for the synthesis of proteins. In the first step, as in every proper cell, the ciphered gene is copied onto the sequence of the mediator RNA. In the next step, this mediator RNA is deciphered, i.e. the RNA is converted from the strange dysfunctional form to the usual functional form, in a process called RNA editing.

Short molecules called **guide RNA** gradually fit themselves to the original RNA molecule from one end and, according to their sequence, the original sequence of the messenger RNA is repaired. The repairs consist in placing nucleotides containing uracil into the correct sites in the original sequence, or these nucleotides are removed from the incorrect sites. This gradually creates the correct form of messenger RNA, which in the third step can then be translated in the standard manner into the amino acid chain of the future proteins and also creates bonding sites to which can be fitted further molecules of guide RNA, capable of directing the editing of an additional

Box 5.7 Messenger RNA

A large portion of genes contain information for synthesis of proteins – long, unbranched chains of amino acids performing practically all the important functions in organisms (e.g. the functions of enzymes, the functions of structural proteins, signal proteins, etc.). However, DNA is not translated directly into the chains of the amino acids forming the proteins, but is first copied into the chains of RNA, molecules that are chemically very similar to a single chain of the double-helix DNA. The protein is formed only in the next step according to this messenger RNA (abbreviated as mRNA). It is the same as when constructing a building, where the construction work is carried out according to a working copy of the original plan stored safely somewhere in an office.

Box 5.8 Deletion of the nucleotide

This is a type of "point mutation". In deletion, a particular nucleotide is removed from the given section. On the other hand, in insertion, a new nucleotide is inserted into a certain position on the DNA. In substitution, one nucleotide is replaced by a different nucleotide. Deletion and insertion of a nucleotide at sites coding proteins usually lead to the loss of the function of the altered protein. This is a result of the fact that translation of the nucleotide sequence in the amino acid chain to the amino acid chain of the protein occurs in triplets, where each set of three nucleotides codes one amino acid of the protein. The insertion or deletion of one nucleotide leads to a frameshift mutation and the almost identical sequence of nucleotides will be translated into a completely different sequence of amino acids. Thisr idiculouss tatementi s ag oodi llustrationo ft hec onsequenceso faf rameshifti nap rintedt ext – an almost unaltered sequence of letters forms a nonsensical sequence of words.

section of the messenger RNA. Frequently, several dozens of different guide RNA molecules gradually participate in editing a gene several hundred pairs of nucleotides long and, in this process, several hundred nucleotides are inserted into or split off from the messenger RNA. The resultant messenger RNA is thus much longer and its sequence differs so much from the original ciphered gene that it is not surprising that, until the RNA editing process was discovered, it was not possible to determine the presence of the particular genes in the mitochondria genome. The molecular apparatus required to edit RNA is very complicated and consists of a large number of components (especially guide RNA molecules). For this reason, a great many scientists searched for the biological function of this process, i.e. looked for the reason why such a complicated apparatus was created in evolution. However, in spite of all the efforts to date, scientists and research workers have not managed to discover the biological function of RNA editing.

Consequently, at the present time, the hypothesis that this is a consequence of passive evolution seems most probable, i.e. that this is a process that gradually developed in some predecessor of trypanosoma as a defense against the consequence of mutations occurring in the original genes, whose products did not require editing. When a mutation occurs in the original gene, e.g. when a nucleotide is deleted at a certain site, which worsens or even abolishes the biological functioning of its protein product (see Box 5.8), evolution can arrange for a remedy in two basic ways:

Either natural selection can fix a mutant, in which the relevant nucleotide is reinserted into the relevant site, or a mutant that is able to somehow neutralize the results of the particular mutation (e.g. in that it is capable of inserting the missing nucleotide into the molecule of the messenger RNA). The second alternative is more probable when there are a greater number of copies of the given gene in the cell or even in the cell organelle. Repair of one copy of a gene by re-inserting the missing nucleotide is generally not substantially manifested in the viability of the organism. However, if the organism forms a molecular apparatus capable of editing all the damaged copies of the given RNA molecule, this can very substantially increase the viability of the individual, which greatly increases the chance that the given manner of renewing the functioning of the gene will be fixed by natural selection. As soon as a species builds up a molecular apparatus that enables repairing of a mutation at the level of the RNA, it will most probably also employ it

Box 5.9 Scientists and research workers

Although the general public has probably not even noticed it, the "scientist" as a special species has been forced out of practically his entire original biotopes in the natural sciences and has been replaced by the much more successful species "research worker". The differences between scientists and research workers are not obvious at first glance. However, research workers do not usually ask "Why?", but rather "How?" and can use complicated and expensive methods to determine which enzyme, which sequence, which molecular weight, which redox potential, or how many molecules of substrate per minute. Average research workers have incomparably greater scientific performance (number of publications and number of citations of these publications) and thus gain higher professional positions than the average scientist. This would be even truer if below-average research workers were compared with below average scientists. To the contrary, the differences will not be so great between top research workers and top scientists. However, because the highly above-average are a negligible minority in the population, for tactical reasons a great many scientists act as if they were research workers and state this profession in their curriculum vitae. For this reason, it is difficult at the present time to determine the exact numbers of scientists and research workers in the scientific community and the first impression that scientists no longer exist in nature could, in fact, be erroneous.

for other mutations. Certain components of the relevant editing apparatus can be utilized universally and can be used to repair other newly formed mutations. Thus, the organism got into a sort of evolutionary trap, a one-way pathway that leads to increasing tolerance for mutations of a certain type in its genes as they are repaired at the level of the RNA. However, this is accompanied by increasing complexity of the editing apparatus. This can even lead to a situation where several dozen, otherwise superfluous guide RNA molecules are required to ensure the functioning of a single gene.

This, by the way, is another example of the shortsightedness of evolution. It is certainly very costly for a species to maintain a complicated editing apparatus. The basic currency in which the price of long-term maintenance of any kind of apparatus in a functioning state in biology consists in the **number of genetic deaths**, i.e. the number of individuals that die (or are not born) because of a damaging mutation in their genes for some component of the particular apparatus. At some time in the distant past, a single genetic death of the first mutant with a deletion would have been sufficient for the relevant gene to remain functional in the particular species. At the present time, the functioning of this gene is dependent on the functioning of dozens of molecules of guide RNA and other components of the editing apparatus. The number of genetic deaths required to ensure maintenance of its function is thus certainly substantially greater than the original number. Thus, if the given model for the formation of editing through passive evolution is correct, then this "discovery" of trypanosoma was almost certainly not worthwhile. (And I am not even mentioning the fact that the existence of a unique and complicated apparatus exposes trypanosoma to the serious danger that our clever research workers will discover an effective chemotherapeutical substance against it. It can be expected that, in time, a drug could be discovered that will prevent the functioning of the editing apparatus and simultaneously will not damage vital functions of organisms without this type of editing apparatus, such as human beings.)

You can't knock a wall down with your head (or your back)

The last Neodarwinist mechanism that could be responsible for the formation of the organization and complexity of organisms is the **wall effect.**[8] The wall effect is manifested as a consistent shift in a single particular direction, which occurs in spite of the complete randomness of the movement of the object in question. If a group of children (or mice or snails) are let go in the middle of an empty room and left for some time, they will move in various directions from their original positions. However, if, before leaving the room we line them up in parallel against one wall, then it is almost certain that they will, on an average, move away from the original wall. This is a result of the fact that the presence of the wall prevents them from moving in one direction and thus they have only the option of moving away from this wall (Fig. 5.4). It is very probable that the wall effect is also important in relation to the organization and complexity of living systems. In order

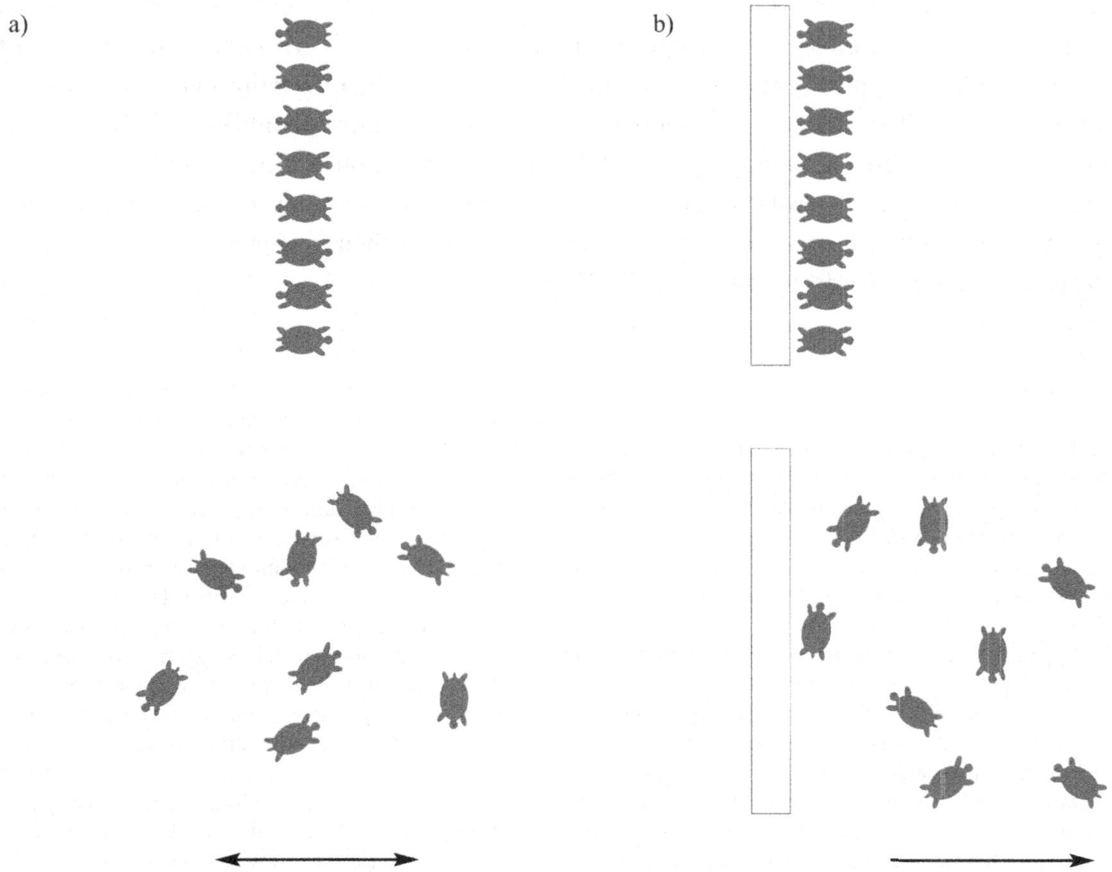

Fig. 5.4 The wall effect. If individuals can move in an arbitrary direction, the population as a whole more or less remains in one place (a). However, if an impermeable barrier prevents movement in a particular direction, the population gradually moves away from it (b). In evolution, a barrier can be created, e.g., by minimum complexity necessary for the functioning of a living system.

to be alive, organisms need to achieve a certain level of organization and complexity. As soon as their complexity or organization decreases below this level, they would no longer be viable and would disappear from nature. At the beginning of evolution, organisms were apparently simple; however, because of the wall effect, their organization and complexity could only increase. Simultaneously, it is probable that, on their pathway through evolution, organisms encountered a number of such walls. There are apparently boundaries below which the organization and complexity of multicellular organisms cannot decrease without them ceasing to be multicellular organisms, boundaries below which the organization and complexity of vertebrates cannot decrease, etc. The existence of these boundaries can gradually or, rather, in a number of steps, direct the evolution of organisms from simplicity to present-day complexity.

Summary and incitement

And, once again, the traditional summarizing of the chapter. A remarkable feature of living systems is their organization, a property that is very frequently erroneously confused with complexity. The complexity and organization of organisms has increased throughout evolution as a consequence of the effect of a number of processes. These include not only Darwinian selection, but also sorting from the standpoint of stability, passive evolution and the wall effect. The next chapter will be concerned with the basic aspect of the formation of new species, i.e. an aspect that was to a substantial degree neglected in Darwin's classical theory of evolution and that was developed only by Neodarwinists in the 1930s.

Footnotes

1. If I recall, this aspect is explained much better (on a much larger area) by Richard Dawkins in his book *Climbing Mount Improbable*. W.W. Norton & Company Ltd. (London), 1996.
2. For those interested in this aspect, I can recommend the book by I. Porigogine and I. Stengers, *Order Out of Chaos*, 1984, Heineman, London.
3. Organisms are, of course, open systems that are far from equilibrium, so that their behaviour (even during ontogenesis) is controlled by the laws of nonequilibrium thermodynamics. However, the structure into which they are arranged is directed, or rather determined (primarily) by the genetic information present in the nuclei of their cells. This means, amongst other things, that this structure is not optimized from the standpoint of maximum dissipation of energy, but rather from the standpoint of efficiency (in K-strategists) or speed (for r-strategists) of transferring of resources to the biomass of the progeny, see Chapter 15.
4. The Deep Hot Biosphere theory assumes that life was formed and continues to exist to a major degree in the depths of the Earth's crust, under quite stable conditions of, from our point of view, high pressures and temperatures, but also far from the reach of detrimental radiation and other unfavourable effects with which organisms on the surface of the Earth must come to terms. The theory assumes that, even today, the majority of biomass is concentrated deep in the Earth's crust and is dependent for energy not on solar radiation, but on the chemical energy. It follows from the theory, amongst other things, that petroleum is derived from stocks of inorganically produced hydrocarbons buried in the depths of the Earth at the time of its formation (and only secondarily contaminated with the products of the metabolism of micro-organisms of deep ecosystems) and that its stocks could be replenished in exhausted deposits (which is said to sometimes occur). I really wish that this theory of Thomas Gold (along with a number of others, at first glance, equally absurd) were true. I can recommend the following article, for example: *Proceedings of the National Academy of Sciences of the United States of America*, 89: 6045–

6049, 1992 and, of course, the book *The Deep Hot Biosphere*. Copernicus Books, 1999.

5. Originally, the possibly more apt term "neutral evolution" was used for this type of evolution of complex traits, see the *Journal of Molecular Evolution* 49: 169–181, 1999. However, because the term "neutral evolution" has long been used in the sense of evolution occurring through the accumulation of selectively neutral mutations, which is a quite different phenomenon, I decided to use the term "passive evolution" instead. In passive evolution, selectively advantageous mutations are accumulated; however, these mutations are advantageous only because they neutralize formerly occurring negative mutations. Thus, evolution functions here as a passive response to an originally detrimental change rather than as the active creation of potentially advantageous new evolutionary traits.
6. A. Stoltzfus described the possible role of neutral evolution in the development of RNA editing in his article in the *Journal of Molecular Evolution* 49: 169–181, 1999. The role of RNA editing in trypanosome (to be more exact in Kinetoplastida) is described in *Molecular Biology Reports* 16: 217–227, 1992 and editing in other systems in *BioEssays* 22: 790–802, 2000.
7. It is certainly worthwhile reading the book by Lynn Margulis concerned with the formation of eukaryotic cells by symbiogenesis. The entire process (especially the formation of flagellum) probably occurred somewhat differently than was originally conceived by the author; however, the scenario was basically similar. *Symbiosis in Cell Evolution. Life and Its Environment on the Early Earth* (San Francisco), 1982, or preferably in the newer *Acquiring Genomes: The Theory of the Origins of the Species*. Basic Books (New York), 2003 or *Symbiotic Planet: A New Look at Evolution*. Basic Books (New York), 2000.
8. The wall effect is described and tested on palaeontological data in the work *Evolution* 48: 1747–1763, 1994 and also in *Scientific American* 271: 63–69, 1994.

Chapter 6 On the formation of species without the participation of natural selection

As I mentioned previously, Darwin's famous book was only marginally concerned with the formation of new species. It was not until the emergence of Neodarwinism in the 1920s and 30s that the aspect of the formation of new species began to be seriously considered. The modern **theory of speciation** – the theory of the formation of daughter species from parent species and multiplication of the number of species in nature – was not established until the 1930s. While Neodarwinists originally placed disproportionate emphasis on the role of natural selection in evolution, they soon concluded that this process generally tends to play a secondary role in the formation of new species. However, the name of this chapter is not entirely precise – the role of natural selection is, nonetheless, important in at least some types of speciation.

Ernst Mayr (1904–2005) and, of geneticists, probably T. Dobzhanski made the greatest contributions to the development of the theory of speciation.[1] Mayr was the first to adequately emphasize the fact that the process of creation of new species occurs not only in time, but also in space. He divided speciation into two basic types, **sympatric speciation**, in which both of the newly formed (mutually differentiating) species are in the same territory at the time of speciation, and **allopatric speciation**, in which they are in different territories or in the same territory but, for some reason, cannot meet.

Sympatric speciation – don't be insulted, neighbour, but I won't reproduce with you

We will first consider sympatric speciation. In this type of speciation, the populations of the original (parent) and newly formed (daughter) species are in contact (Fig. 6.1). As a consequence, they affect one another ecologically in that they compete for the same resources and, in addition, their members can cross. Crossbreeds of the parent and daughter species, of course, wipe out the genetic differences between the two species and thus reduce the probability of their mutual diversification. Amongst other things, this reduction of the differences prevents the two species from dividing the ecological niche so that each of them could specialize in using a certain part of the resources. In the absence of such specialization, i.e. without differentiation of the ecological niches (see Box 2.10 on p. 21), however, the two species cannot continue to exist in a single territory. The competitively stronger species will sooner or later force out the competitively weaker species, so that any speciation would not lead to an increase in the number of species – either the new species would replace the older species or would disappear after some time.

Natural selection can play an important role in certain forms of sympatric speciation.

For example, natural selection is the most important factor in **ecological speciation.**[2] This type of sympatric speciation can occur if the environment of the species contains two very different biotopes. In this case, disruptive selection acts on the species, i.e. selection favoring two specialized forms of a particular species and simultaneously placing at a disadvantage individuals with a transitional phenotype that are thus not ideally adapted to either of the biotopes.

Consequently, crosses between the two specialized forms are at a disadvantage and thus a selective pressure systematically acts on the species towards the formation of some sort of mechanism preventing their crossing.

In nature, a number of pairs of species are known that were apparently formed through the action of ecological speciation. Primarily, this applies to the historically documented case of speciation of the apple maggot fly *Rhagoletis pomonella* in America, where part of the population moved from the hawthorn to apple trees in the middle of the 19th century.[3] Hawthorns and apple trees were present in the same territory. Partial reproduction isolation of the original and newly

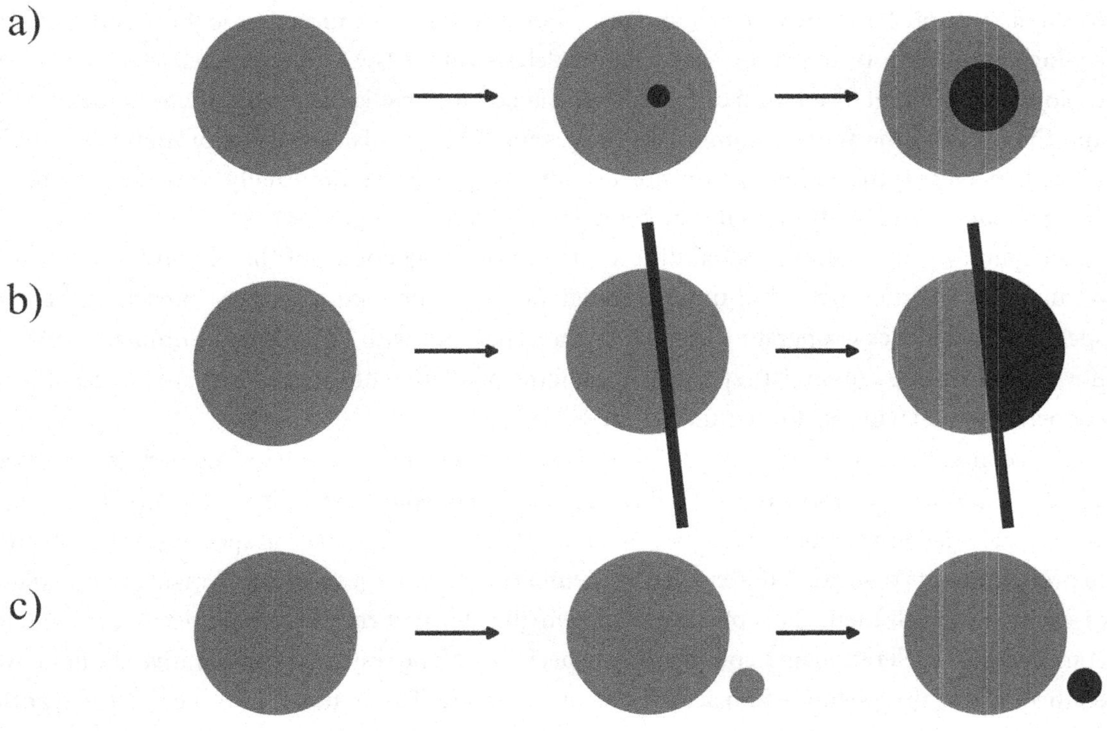

Fig. 6.1 Basic types of speciation. Sympatric speciation (a) means the formation of a new species within the area occupied by the parent species. The members of the old and new species can meet during sympatric speciation. Vicariant allopatric speciation (b) means the formation of a new species as a consequence of division of the area occupied by the original species into two or more parts of comparable size as a consequence of the formation of geographic barriers (e.g. a new river). Peripatric allopatric speciation (c) is the formation of a new species as a consequence of splitting off of a very small part of the parent population, e.g. by colonization of an island far from the mainland.

Box 6.1 Prezygotic and postzygotic reproductive isolation barriers

Prezygotic reproductive isolation barriers are all the barriers preventing crossing between the members of two different species that are active before combination of the male and female sex cells and thus prior to the formation of the zygotes (germ cells). Barriers can consist, for example, in rejection of members of the second species as potential sexual partners, in the incompatibility of the patterns of behaviour that, under normal conditions, precede mating, in mechanical incompatibility of the male and female sex organs, in the inability of the male sex cells of one species to survive in the female sex organs of the second species or in the inability of the male sex cells to seek out female sex cells and to combine with them. Postzygotic reproductive isolation barriers are all the barriers preventing development of the fertilized sex cells formed by inter-species crossing into viable and simultaneously fertile progeny. While prezygotic reproductive isolation barriers can be formed by the action of natural selection, as their formation prevents the unnecessary investment of resources into inter-species crosses exhibiting worse viability or fertility, postzygotic reproductive isolation barriers are generally formed by gradual accumulation of random mutually incompatible changes in the gene pools of the two species.

formed species of fly was apparently initially ensured by the different time of ripening of the fruit of the apple and hawthorn, in which these flies developed. Another probable candidate for **ecological speciation** consists in American sticklebacks of the *Gasterosteus* genus, which form two ecologically different forms with different appearances in Canadian lakes, one living at the bottom and the other in the water column. It is interesting that, in this case, pairs of mutually closely related species in the same lake do not usually cross; however, mutually unrelated forms in different lakes can readily cross if they belong to the same ecological form.[4]

In sympatric speciation, partial differentiation of the genomes of the old and new species occurs after a certain period of time. Mutated alleles accumulate separately in each of the two species; these alleles cooperate well with one another and with the original unmutated alleles; however, if, during crossing, they enter the genome of a different species, they are not capable of cooperating well with the local mutated alleles.

As a consequence of this incompatibility of the mutated alleles of one species with the mutated alleles of another species, crosses of the two species have reduced viability or fertility[5] (Fig. 6.2).

Natural selection can once again become important in this late stage of speciation. This begins to prefer mutants that can differentiate the members of their own species and mate preferentially or exclusively with them. This phenomenon, which is termed **reinforcement**, can contribute to considerable acceleration and completion of speciation.[6] The results of comparative studies have confirmed that this phenomenon actually occurs in nature. These studies indicate that **prezygotic**

Box 6.2 Biotope

A biotope is a type of natural habitat that is characterized by the species composition of the organisms living in it, and the amount, type, and the place and time of occurrence of all resources. It is further characterized by typical physical, chemical and biological factors that can affect the lives of its inhabitants. Typical biotopes are forests, steppes or the bottom of the ocean.

Box 6.3 Gene pool

The gene pool is the sum of all the alleles (gene variants) occurring in the members of a certain population. The gene pool of a species is the sum of all the alleles occurring in the members of a certain species.

reproductive isolation barriers occur much more frequently in sympatrically occurring species, preventing crossing of the two species, e.g. in that they prevent copulation of members of the new and old species. For example, these barriers can consist of differences in the ways in which the males court the females. In contrast, in species that developed in different territories (allopatric) and were thus not exposed by selection to the ability to differentiate the members of their own species, prezygotic reproductive isolation barriers are much rarer and **postzygotic reproductive isolation barriers** are far more common – for example, the formation of unviable or infertile crosses.[7]

Theoretical analyses and the results of laboratory experiments on fruit flies have, however, demonstrated that prezygotic reproduction isolation mechanisms are not readily formed by the reinforcement mechanism. Unless a very strong reproductive isolation barrier is established

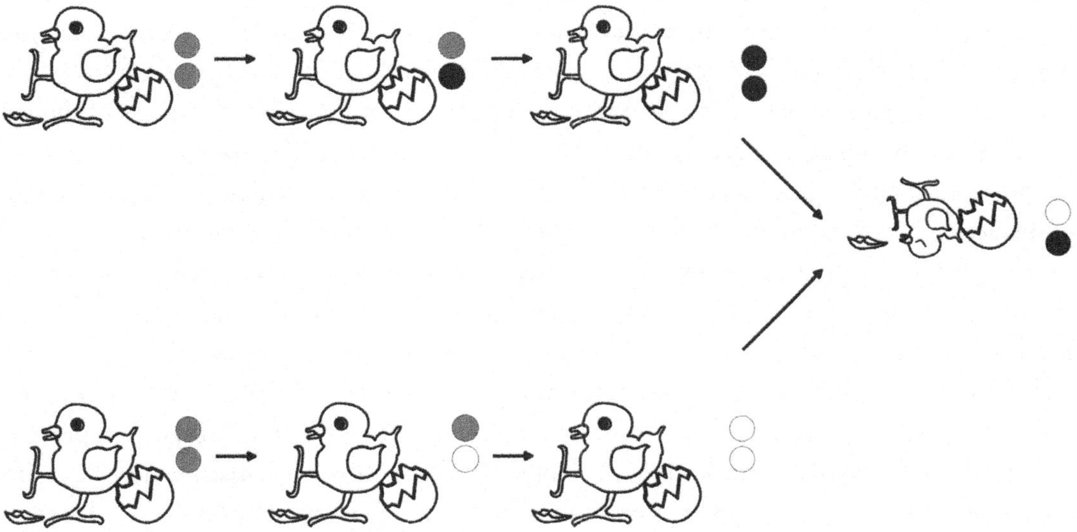

Fig. 6.2 The formation of a new species as a consequence of fixation of mutually incompatible alleles in two isolated populations. Initially, the members of the two isolated populations have the same alleles (grey circles). After some time, new alleles are formed in each of them, black circles in one population and white circles in the other. Both new alleles are compatible with the original alleles, i.e. heterozygote individuals carrying one old (grey) and one new (black or white) allele are viable. In time, the new alleles will predominate in each of the populations (as a consequence of selection or drift). If individuals of both populations subsequently cross, the crosses will carry two different new alleles; if these alleles are mutually incompatible, it will not be viable.

between members of the two extreme forms, the gene for the characteristic on the basis of which the members of the same form recognize suitable sexual partners or the gene for preferring this characteristic can, at any time, pass through crosses to the gene pool of the wrong form. This understandably greatly complicates the establishment of reproductive isolation between the members of the two forms. Some biologists are of the opinion that, for this reason, cases of ecological speciation are relatively rare.

Another type of sympatric speciation consists in **ethological speciation**. The theory of ethological speciation assumes that the first step leading to the formation of a new species is mutation resulting in a change in the behaviour of the mutants. As a consequence of the change in behaviour, the bearers of the relevant mutation begin to preferentially breed together. The change in behaviour can consist, for example, in seeking out different places for breeding or in a change in the way in which an individual recognizes members of the opposite sex of its own species. As soon as sufficiently strong reproductive isolation barriers are formed through the effect of the change in behaviour, the individuals of both (now genetically isolated) populations obtain the chance and primarily the time to differentiate in their phenotypes (e.g. body structure) and, in time, also ecologically (e.g. in food habits).[8]

Allopatric speciation – my mother is crying away over the hills...

Now we have arrived at **allopatric speciation**. This type of speciation probably occurs more frequently in nature than sympatric speciation. Allopatric speciation leads to the formation of a new species in a population that is not in direct contact with other populations of the parent species. Thus, ecological and genetic interactions do not occur between members of the parent and the newly formed daughter species, or this occurs to only a limited degree and the whole process of speciation is easier. On the other hand, if the old and new species later come into contact, they need not yet have sufficiently strong reproductive isolation barriers, can cross together and, if the viability or fertility of the crosses is not substantially reduced, can once again merge into a single species. However, if the time of genetic isolation of the two species is suitably long, a sufficient number of incompatible mutations is accumulated in the gene pools of the two species, so that adequate genetic isolation of the two species is finally provided by postzygotic reproductive isolation barriers consisting in the lack of viability or infertility of any crossbreeds. The results of experiments and observations in nature have shown that, when the two populations are exposed to substantially different selection pressures, mutually incompatible mutations can accumulate faster in their gene pools and a substantially shorter time can be sufficient for the formation of internal reproduction barriers.[9]

In domestic animals, the breeder ensures reproduction isolation of the members of new breeds in that he decides which individuals are to breed together. As a consequence, large differences between the appearances of various breeds can be achieved in a small number of generations; however, reproduction isolation barriers are not formed simultaneously. There is no selective pressure on the formation of prezygotic reproduction barriers and far more time would be required for the formation of postzygotic reproduction barriers as a consequence of accumulation of incompatible mutations. Thus, a dachshund and St. Bernard dog remain members of the same species and can breed successfully. In the absence of constant intervention on the part of human beings, all the breeds of dogs would gradually merge into a single breed – the universal mongrel. Alright, I'll admit it – this wouldn't be very easy for St. Bernard dogs and dachshunds, but through at least one or two generations of intermediaries, the genes of both breeds could bark in the form of a single cute puppy.

Ernst Mayr differentiated two basic types of allopatric speciation, vicariant (dichopatric) speciation and peripatric speciation. In **vicariant speciation**, the original population of the parent species is most frequently divided by a newly formed natural barrier (river, mountain range or, for marine organisms, a strip of land) into two new populations of comparable size. In **peripatric speciation**, a small daughter population breaks off from the parent population and, in time, forms the basis for the formation of a new species. Small daughter populations are probably very frequently split off from most species. A group of individuals (or even the legendary fertilized female) can, for example, successfully settle on an island sufficiently far away from the mainland. Small isolated populations are probably formed far more frequently on the mobile borders of the occurrence of the species. As meteorological conditions change over time, the area of occurrence of the species becomes larger and then smaller or moves to a different geographic area. At suitable places along the borders of the area, small populations remain and can, in time, become isolated from the main area of occurrence of the relevant species by extensive territories where the conditions are unfavourable for the particular species. Most of these populations disappear in time when the area of occurrence of the main population expands again and encompasses the area of occurrence of the local population, or the local population dies out. In some cases (apparently very rare, but evolution has plenty of time for repeating unsuccessful experiments), the local population develops into a new species, which can survive permanently at the given site or can gradually expand, sometimes at the expense of the area of occurrence of the main population. If isolation lasted for a sufficiently long time and sufficiently strong reproduction isolation barriers were formed between the new and old species and if the new species is better adapted to the local conditions, it can even force out the old species and occupy its niche in the ecosystem. If the ecological requirements of the two species are differentiated (their ecological niches become different), then both species can survive in the same place for a long time. Mayr and a number of other authors are of the opinion that peripatric speciation is accompanied by significant genotype and thus phenotype changes in the newly forming species. In vicariant speciation, the two populations have very similar gene pools and thus the two resultant species do not differ much. However, in peripatric speciation, the new species takes with it only a small part of the genetic variability of the original species. Which alleles become part of the gene pool of the new species is substantially a matter of chance. In this connection, Mayr even spoke of a **genetic**

revolution, i.e. a drastic change in the genetic composition of the new population.[10] Genetic studies have actually demonstrated not only that most of the rare gene variants (rare alleles) are lacking in the new species, but also that some alleles that were originally very common are missing or are present only minimally. On the other hand, some originally very rare alleles occur in unusually high numbers. Thus the importance of peripatric speciation could lie in the fact that it can be accompanied by drastic changes in the genotype, phenotype and thus in the ecological requirements of the newly forming species. Even if vicariant speciation were to occur more frequently than peripatric speciation (which I very much doubt), peripatric speciation would still be of greater importance in diversification of the species.

Reproduction isolation while you wait, speciation in a single day

In addition to typical cases of gradual sympatric and allopatric speciation, which usually take a very long time, cases of **instant speciation** are also known. **Polyploidization** – the multiplication (most frequently doubling) of the number of chromosome sets in the cells of the mutant – is an example of such instant speciation. Especially in plants, polyploids are very frequently viable and can even cross together. However, if a tetraploid (i.e. an individual with four chromosome sets in its cells) crosses with a normal diploid (i.e. an individual with only two chromosome sets), then triploids are formed, which are very frequently incapable of multiplying, even if crossed among themselves. In meiosis, one of the three homologous chromosomes does not have a partner with which it could pair and recombine (meiosis and homologous chromosomes were described in Chapter 3). Unpaired chromosomes then freely "wander around" in the cell and prevent the completion of cell division. In some species, however, triploids can multiply permanently without sexual reproduction, for example, by tillering or through rhizomes and tubers, and thus a separate species can be formed. The fact that a polyploid cannot cross with the original diploid species leads to the instantaneous formation of a perfect reproduction barrier that is capable of ensuring the separate existence of the new species and thus provides it time, in the presence of the original species, to differentiate from the original species, both in appearance and ecologically. In addition, doubling of the amount of genetic information is generally reflected in the appearance of the newly formed species. For example, in plants, tetraploids are usually larger than diploids. In addition, apparently because they can have four variants of each gene (each potentially suitable for a different situation), they are usually more resistant to unfavourable effects and are capable of

Box 6.4 Karyotype

The term karyotype designates a species-specific set of chromosomes in a cell. A karyotype is characterized by the number and shape of all the chromosomes in the nucleus. It can be studied in the time intervals of division of the cell nucleus when the individual chromosomes are most highly condensed and form conspicuous rod-like shapes that can be studied when suitably dyed under an optical microscope.

surviving in places with extreme conditions. This again assists in mutual ecological diversification of the old diploid and new tetraploid species and gradually permits the common occurrence of both species in the same territory.

Different kinds of karyotype changes, i.e. changes in the number and structure of the chromosomes other than multiplication of the entire genetic set, can also play a role in the formation of a new species. An entire long chromosome section can move from one chromosome to another, a chromosome can divide into two smaller ones or, on the other hand, two small chromosomes can merge to form one large one. This can (but need not) reduce the probability of recombination between the new and the old forms of the chromosomes which can assist in speciation. Under certain conditions (mainly in heterozygotes), karyotype changes can cause disorders in cell division and thus reduce the fertility of crosses of the original and mutated forms. In fact, some authors think that most speciation is accompanied by or is even caused by chromosome changes creating reproductive isolation barriers at the very beginning of the formation of the species. In contrast, other authors are of the opinion that the vast majority of karyotype changes observed in nature are, in actual fact, the consequence and not the cause of speciation.[11] While most newly formed karyotype changes substantially reduce the viability or fertility of individuals, the successful ones, i.e. those that we encounter in a greater number of individuals in nature, have already passed through the imaginary sieve of natural selection and thus have demonstrated, amongst other things, that they do not substantially reduce the viability of crosses with the bearers of the original, and thus initially more frequent, karyotype forms. As a consequence, these (successful) karyotype changes cannot form substantial reproduction barriers that would lead to the splitting off of a daughter species.[12]

Thus, karyotype changes cannot apparently be the main driving force in speciation; however, they can be a very common product of this process. Individual types of chromosome changes leading to a change in karyotype occur relatively frequently in the population. The chromosome mutations can then compete in the area of "**meiotic drive**". Meiotic drive is an evolutionary mechanism that can lead to very fast spreading of a new mutation in the population. Some mutations, including chromosome mutations, are capable of affecting the process of meiosis in a way that increases the probability that they will be present in the functional products of meiosis, i.e. in functional sex cells, and thus subsequently in the gene pool of the next generation. For

Box 6.5 Oocyte and polar body

An oocyte is an immature female sex cell. In contrast to male sex cells, an oocyte does not divide into four identical sex cells during meiotic division, but retains practically all its original cytoplasm and other cellular content. In each of the pair of divisions that together form meiosis, the future egg retains only one nucleus and the other is forced to its surface as part of the polar body. The polar bodies and all the DNA that it contains are excluded from the oocyte and thus do not contribute to the formation of the embryo.

Box 6.6 Chromosome and chromosome mutation

Each chromosome is formed of two identical chromatids, where each chromatid is formed of two arms separated by a centromere – a narrower point where the two chromatids are held together until a certain instant in nuclear division by unreplicated DNA (and simultaneously are held together at other points by special proteins). If a centromere is located in the chromosome close to the end of the chromatid, the chromosome apparently has only one arm. Examples of chromosome mutation are the division of a large chromosome into two small ones or the fusion (merging) of two small chromosomes into one large one. The best known example of chromosome mutation is probably Robertsonian translocation (centric-fusion), in which two chromosomes with centromeres close to their ends (acrocentric chromosomes) fuse to form one chromosome with a centromere close to the centre (metacentric chromosome). A heterozygote individual has a genome containing both the pair of original acrocentric chromosomes and also a new metacentric chromosome. During meiosis, this set of three chromosomes forms a "trivalent" and, at the end of the nuclear division (if this turns out well), the metacentric chromosome goes to one nucleus and the two acrocentric chromosomes go to the other nucleus. In some species, the metacentric chromosome ends up with greater probability in the pole body (i.e. a body that later "dies" together with its chromosomes); in others, the acrocentric chromosome ends up in the polar body.

example, the variant of a chromosome that is capable of entering the nucleus of an egg cell with greater probability than a polar (residual) body (that, in contrast to the egg cell, is predestined to die) can very rapidly replace the original variant of the particular chromosome.[13]

It is interesting that the genetic composition of the population need not be changed during this process, i.e. the contents of the individual alleles of the genes present on the altered chromosomes remain constant. In crosses, the old and new form of the chromosome can recombine and exchange alleles in at least some places.[14] In these cases, only the basic shape of the chromosome is inherited from generation to generation, while its content (the set of alleles that it contains) is not.

The reason why most speciation seems to be accompanied by karyotype changes probably lies in the fact that, after the formation of reproduction barriers between the new and old species, competition begins to occur amongst the individual variants of the chromosomes isolated in the populations of the two species. Thus, other variants of chromosomes spread very rapidly in each of the species through meiotic drive and the karyotypes of the two species then differ (Fig. 6.3).

Speciation without sex – anyone can do that

At the end of the chapter, at least brief mention should be made of the aspects of **speciation in species without sexual reproduction**, called **asexual species**. This aspect of speciation has been treated in far less detail. Until recently, it was not even clear what holds the species together here – i.e. what makes it possible to more or less differentiate amongst various species, between which there are not many transitions. In sexual species, the similarity of the members of a single species is ensured by cross-breeding and the related exchange of genetic information. However, this factor is understandably missing in asexual species. One of the processes that could replace this is natural selection. If each asexually multiplying species were optimally adapted to a particular type of environment, all the individuals with transition properties could be removed by natural selection. Newer theories assume that **evolutionary hitchhiking** (genetic draft) could play an important role in maintenance of species differentiation in asexually reproducing organisms, see Chapter 4. In asexually multiplying organisms, the evolutionary fate of all the genes in a single genome is totally connected. The processes of segregation and recombination do not occur here and thus there is an absolute **gene linkage** amongst all the genes (see Box 4.10 on p. 59). Thus, if a very advantageous mutation appears from time to time in the genome of an individual in the asexual population, not only does it spread at the expense of unmutated copies of the relevant gene, but it also removes all the genetic variability in the other genes present in the population. After a certain time, only copies of the mutant chromosomes remain in the population. **According to this conception, occasional waves of elimination of genetic variability accompanying the spreading of advantageous mutations are responsible for the mutual genotype and phenotype similarity of the members of asexually multiplying species.**[15]

Thus, the situation is the opposite for asexual and sexual species. We would expect that asexual species would continually split into ever greater numbers of new species. As this doesn't happen in nature, we are glad when we find a process that prevents continuous speciation (such as a selective sweep). These mechanisms are apparently more effective than they seem at first sight. Biodiversity amongst asexual organisms is substantially less than amongst sexual organisms. Possible explanations for this phenomenon will be discussed in Chapter 15.

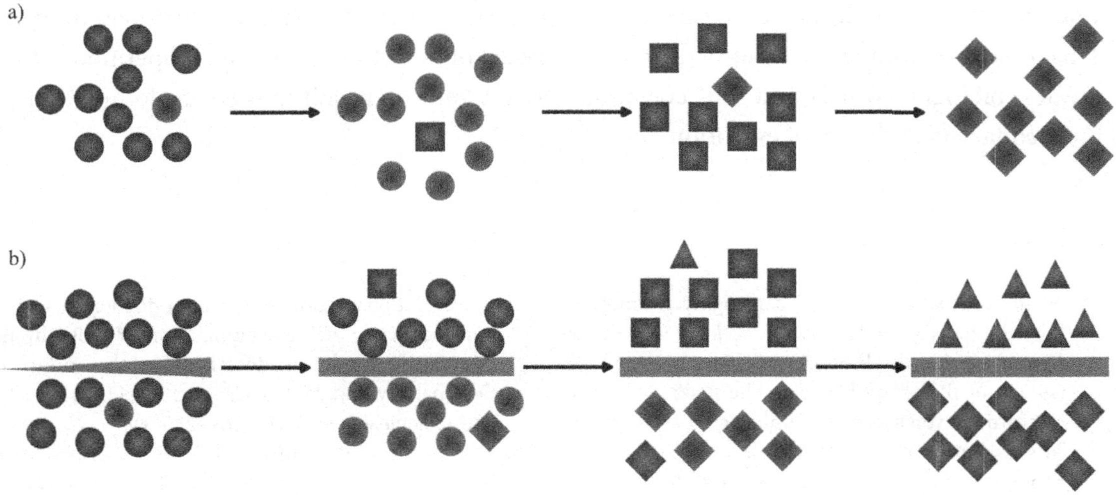

Fig. 6.3 Diversification of chromosomal sets in two isolated populations through meiotic drive. Meiotic drive is usually much faster than speciation. Thus, changes in the chromosome sets (i.e. karyotypes) occur during the existence of the species. If the population forms a single unit (a), rapid fixation occurs in all the individuals of the given species, so that we do not learn of changes in the karyotype. However, as soon as speciation (b) occurs in the species, the population divides into two parts and, from this moment, different karyotypes are fixed in the two parts by meiotic drive.

Summary and incitement

To summarize. In his famous book "On the Origin of the Species by means of Natural Selection", Darwin did not spend much consideration on the mechanism of the formation of new species However, it is apparent that he was of the opinion that natural selection played a key role in the formation of new species. The modern theory of speciation was established only in the 1930s, mainly through the work of Ernst Mayr. He demonstrated that the main problem in the formation of a new species amongst sexually multiplying species lies in the formation of sufficiently strong reproductive barriers capable of preventing exchange of genes and thus eroding the differences between the newly forming species. He demonstrated that the main role is generally played by allopatric speciation, i.e. the differentiation of species in various territories (and thus out of mutual contact). In vicariant speciation, the population of the original species is divided by new natural obstacles, such as a river, into two populations of approximately the same size, as a consequence of which the genetic composition of the two new species does not differ much from that of the original species. In the second type of allopatric speciation, peripatric speciation, only a few individuals split off from the large population and take with them only a very small part of the gene variants originally present in the large population. Sympatric speciation, i.e. speciation occurring in a single territory, is also possible but is substantially less common. According to Neodarwinists, natural selection played a much smaller role in the formation of new species than that attributed by Darwin himself.

The next chapter will be concerned with the subject of frequency-dependent selection, the study of which led at the end of the 20th century to the discovery of evolutionarily stable strategies. This discovery limited the validity of Darwin's model of evolution based on competition for the greatest biological fitness and thus constituted a very important jolt to the very foundations of the Neodarwinist theory of evolution.

Footnotes

1. Of the books by E. Mayr, I can highly recommend the two books Mayr, E. *Animal Species and Evolution.* Harvard University Press, Cambridge, 1964 and *The Growth of Biological Thought.* The Belknap Press of Harvard University Press, Cambridge, 1982. Note the latter is, however, rather extensive.
2. Ecological speciation is described, e.g., in the article *Trends in Ecology & Evolution* 16: 372–380, 2001.
3. A summary article on ecological speciation in the maggot fly can be found, e.g., in *Nature* 407: 739–742, 2000.
4. Speciation of sticklebacks is described, e.g., in *Trends in Ecology & Evolution* 13: 502–506, 1998, *Evolution* 56: 1199-1216, 2002, *Trends in Ecology & Evolution* 17: 480–488, 2002, *Trends in Ecology & Evolution* 19: 456–459, 2004.
5. The evolutionary formation of reproduction barriers as a consequence of the accumulation of mutually incompatible alleles is described in the model of Dobzhansky and Muller *Genetics* 21, 113–135, 1936 and *Biological Reviews* 14: 261–280, 1939.
6. The process of reinforcement is described, e.g., in *Theoretical Biology* 160: 163–174, 1994, *Nature* 387: 551–553, 1997 and *PLoS Biology* 2: 2256–2263, 2004.
7. The differences in the strengths of prezygotic and postzygotic reproduction barriers in sympatric and allopatric pairs of species of fruit flies are described in the work *Evolution* 43: 362–381, 1989 and the absence of prezygotic barriers in allopatric species of Darwin's

finches is described in the work *American Naturalist* 160: 1–19, 2002.

8. The possibility of ecological speciation, especially as a consequence of changes in the recognition of sex partners, are discussed, e.g., in *Nature* 411: 944–948, 2001, *Molecular Ecology* 10: 1075–1086, 2001, *Trends in Ecology & Evolution* 16: 364–371, 2001.
9. If the same species occupies two different environments, the formation of reproductive isolation can be very fast. For example, a study performed on salmon *Oncorhynchus nerka* demonstrated that a reproductive isolation barrier was formed between populations introduced in a river and in a lake within 13 generations, see *Science* 290: 516–518, 2000.
10. The phenomenon of genetic revolution is described very nicely in Mayr's book, *Animal Species and Evolution.* Harvard University Press, Cambridge, 1964. Theoretical models are described in the work *American Naturalist* 147: 466–491, 1996.
11. The relationship between speciation and karyotype changes (in short, everything you ever wanted to know about karyotype evolution and were afraid to ask...) is described in the book by M. King (1993), *Species Evolution. The Role of Chromosome Change.* Cambridge University Press, Cambridge.
12. The chromosome mutations that are encountered in the laboratory and in nature are very different in some respects. Mutations observed in the laboratory have not yet gone through the sieve of natural selection and thus the vast majority of these reduce the biological fitness of their carriers, or at least the fertility of crosses carrying the mutated and unmutated forms of a certain chromosome in the nucleus. In contrast, in the natural population, we quite frequently encounter chromosome mutations that do not affect the viability or fertility of their carriers. They have already passed through the sieve of natural selection and have been shown not to be detrimental. (The probability that an undetrimental and thus, subsequently, more common chromosome variant will be encountered in nature is understandably greater than that we will encounter a detrimental and thus rare variant.) In some species, there are a number of chromosome races that differ (frequently only) in their karyotype. This situation readily occurs when the karyotypes of the individual newly formed races are compatible with the initial karyotype (consequently, they can initially spread in a population consisting of the carriers of the original karyotype); however, they are not mutually compatible – their crosses are not fertile.
13. A specific example of this type of meiotic drive is described, e.g., in *Mammalian Genome* 6: 315–320, 1995 (and a great many other works) and a number of examples are given in King's monograph – see Note 11.
14. If the crosses of two chromosomal races have reduced biological fitness, then the members of one chromosomal race have great difficulty in penetrating more deeply into the area of occurrence of the other race. Any immigrant must cross primarily with the members of the local population, so that it has overall low fitness. However, if this fitness does not equal zero, even a small number of immigrants can bring many of their alleles into the foreign population, as a consequence of which the chromosome races cannot be differentiated into independent species (with different phenotypes).
15. The role of genetic draft in maintenance of species cohesion in asexual organisms is described, e.g., in *Annual Review of Microbiology* 56: 457–487, 2002.

CHAPTER 7 How Darwinism survived its own death – frequency-dependent selection and the theory of evolutionarily stable strategies

Darwin never stated that natural selection would be the only driving force for biological evolution. The discovery of genetic drift, evolutionary hitchhiking, sorting from the standpoint of stability and passive evolution would thus not have disturbed him at all and, in fact, might have pleased him. Neodarwinists, who substantially preferred the role of natural selection, understandably had to come to terms with a number of unpleasant facts as time progressed. For example, molecular biologists and geneticists caused them great difficulties with their **theory of neutral evolution**[1], which assumes that most traits were fixed during evolution without the action of natural selection – see Chapter 4. Nonetheless, taken objectively, none of the discussed, newly described phenomena endangered the foundations of the theory of evolution. The aspect of evolutionarily stable strategy, which was developed in the 1970s by John Maynard Smith and George Price, belongs in a category that caused greater difficulties for the theory of evolution, although it finally managed to encompass this too.

About Hansel and Gretel and frequency-dependent selection

Natural selection occurs in a number of forms in nature. **Frequency-dependent selection** is a very important and widespread type of selection. In this form of selection, the biological fitness of the bearers of a certain trait depends on their fraction in the population. In some cases, the biological fitness of the carriers of a trait increases with their increasing frequency in the population, while it decreases with their increasing frequency in the population in other cases.

Copying the choice of a sexual partner is an example of **positive dependence of biological fitness** of the bearers of a certain trait on their frequency in the population. It is known that the females of many species of fish, birds and mammals mate preferentially with males that are preferred by most of the other females in the population. Thus, a female observes which males are most attractive for the greatest number of females and then, perhaps contrary to her own taste, copies the majority taste of the population. For example, research workers studied which of two males would be preferred by a female of the live-bearing guppy fish *Poecilia reticulata* who had previously observed that another female preferred the male with less bright colouring.[2] Under normal conditions, females always prefer the more orange male. However, when the female saw that another female preferred a less orange male (actually, she had no choice, because the research workers separated the more orange male with a glass plate), in the subsequent experiment she began to substantially prefer the less brightly coloured male as a sexual partner (Fig. 7.1). From an evolutionary standpoint, this conformist behaviour of the females makes a certain kind of sense. Sons thus have a greater chance that they will be favoured by a large number of females even if their mother has wandered into another population where a different "fashion" is predominant. However, this simultaneously leads to the traits of males being subject to positive frequency-dependent selection. The selection value of their appearance (i.e. of their phenotype) will depend on how frequently this phenotype occurs in males in the population. If this is a common phenotype, the average female will frequently see that other females mate with the bearers of this phenotype. Thus, she will prefer the bearers of this phenotype compared to the bearers of less

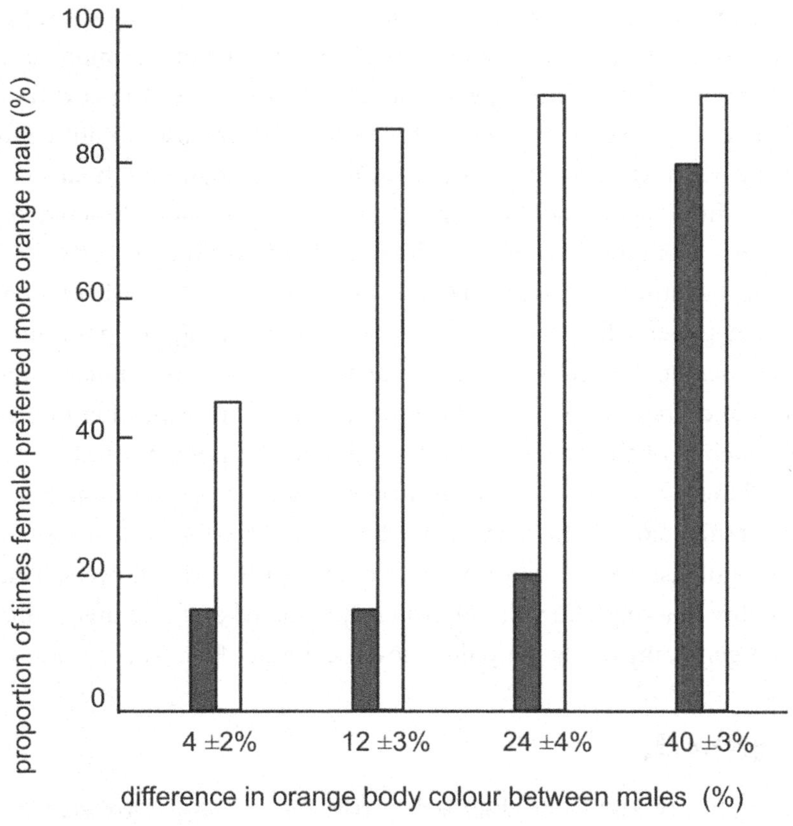

Fig. 7.1 Female preference for preferred guppy males. Under normal circumstances, guppy *Poecilia reticulata* females prefer males with a larger area of orange colouring. In the experiment, females were allowed to choose between males with different areas of orange colouring on the surface of their bodies under normal circumstances (white column) or after they saw a less orange male courting a different female (black column). The results show that, provided the difference in the orange colouring is not very great (about 40%), females greatly prefer males that attracted the interest of other females, in spite of the smaller area of orange colouring on their body. The height of the columns shows the percentage of cases in twenty independent experiments in which the female preferred the more orange male.

common phenotypes. Thus, the greater the frequency of a certain phenotype in the population, the greater will be the biological fitness of its bearers and the more progeny they will have during their lifetime. If no other force acting in the opposite direction is simultaneously active, positive frequency-dependent selection will finally lead to the fixing of a certain phenotype in the population, while the other phenotypes will be removed by selection.[3]

A different situation occurs in connection with **negative frequency-dependent selection.** In this situation, the biological fitness of the carriers of a certain trait decreases with its increasing frequency in the population. An example could consist in some kinds of orchids (for example the elder-flowered orchid *Dactylorhiza sambucina*) in whose population two coloured forms – yellow and purple – are permanently found. Externally, both forms are very attractive for their insect pollinators; however, neither form contains nectar in its flowers. A bumble bee flies to a purple flower and searches for nectar, but can't find any anywhere and finally, disappointed, flies to a different plant. As can be expected and as has been found by observations in nature, bees far more frequently prefer yellow flowers over other purple flowers for the next try. However, even here the bee is not very successful, because the yellow flowers also do not contain any nectar. Consequently, the bee next tries purple flowers again and so on, until it finally occurs to them to seek out a better kind of plant that does not have such beautiful and attractive flowers, but does have nectar in them. Nonetheless, it is quite possible that next time the bee will again let himself be attracted by the beautiful but false gold of the orchid flowers.[4] As marketing professionals well know, good quality goods sell themselves, however only assuming that poor quality goods don't have better advertisements. Negative frequency-dependent selection, similar to the situation with orchids, often leads to the permanent occurrence of several forms together in the population. As soon as the frequency of the yellow form decreases in the population, for example if Gretel convinces Hansel to pick a yellow bouquet of strictly protected orchids (together with the bulbs), the rare bearers of this color will then have a greater chance that they will be visited by a pollinator after an unsuccessful visit to the more common orchid with purple flowers. Thus, there will be a greater probability that the yellow flowers will be pollinated and a greater probability that they will pass on their genes, including the yellow flower color, to the next generation. Because of the

Box 7.1 Feedback

Feedback is a concept from cybernetics. This is basically the direct or mediated action of a signal from the output of a certain element to the input of the same element. Thus, positive feedback occurs when an increase in the value of the output signal brings a signal to the input of the same element that leads to an increase in the output signal (heating burning material increases the intensity of burning and this again increases the heating of the burning material, which increases the intensity of burning, etc.). Negative feedback is involved when an increase in the value of the output signal brings a signal to the input of the same element that leads to weakening of the output signal – see, for example orchids. Negative feedback forms the basis of all regulation – it allows the system to be maintained in an equilibrium and thus in a stable state even under conditions where it is exposed to random disturbing effects. Such as Hansel and Gretel pulling up rare orchids.

Box 7.2 Frequency-dependent selection and left-handed people

According to some authors, the frequency of left-handed people in the population is regulated by frequency-dependent selection. Left-handed people are at an advantage in battles and fist-fights (and also in a number of sports) as their opponents are not prepared for their fighting methods because there are fewer of them in the population than right-handed people. Comparative studies on a large number of traditional human populations have actually shown that that the frequency of left-handed people (3–27%) is directly dependent on the amount of violence in the population (specifically, the number of violent deaths).[5]

action of this negative feedback, the relative number of yellow orchids will increase somewhat in the next generation and the relative number of purple orchids will decrease.

After a few generations, the ratio of the purple and yellow orchids in the meadow will return to the original value. Negative frequency-dependent selection is very extensive in nature and is apparently to a great degree responsible for the simultaneous occurrence of several forms of a certain trait in natural populations. Positive frequency-dependent selection is apparently less common. As systems containing positive feedback (a decrease causes a further decrease, an increase causes a further increase) are, in principle, unstable (because, as is well known, no tree grows up to heaven and pride comes before a fall), it must necessarily be encountered less frequently in nature than systems with negative feedback or without feedback.

By the way, the proponents and frequently even the opponents of the Gaia theory are not always aware of this simple fact. (I certainly don't suspect James Lovelock, the spiritual father of the Gaia theory, of anything of the sort.)

Advocates of this theory frequently remark on how ingeniously the global system of the Earth functions and how it has created multifarious mechanisms for maintaining the temperature and chemical composition of the atmosphere and hydrosphere within a narrow range of values

Box 7.3 Gaia

The name of the ancient Greek goddess Gaia is used to denote the hypothetical superorganism consisting of the entire biosphere of the planet Earth. As the presence of various homeostatic mechanisms is typical for organisms, maintaining their individual body parameters (temperature, chemical composition, etc.) within the physiological boundaries required for the life of the organism, similarly, in the biosphere of the Earth, we can find a great many regulation mechanisms maintaining various physical and chemical conditions predominating on the Earth within boundaries compatible with survival of the individual species. It is only thanks to the activities of living organisms that, for example, the temperature of the surface of the Earth remains constant over long periods of time, regardless of the fact that the amount of light from the sun and thus the input of energy has increased substantially over the past 4 billion years. Similarly, the activities of organisms regulate the level of oxygen and carbon dioxide in the atmosphere. In the absence of life, the conditions on Earth would have become similar to those on Mars long ago. The activities of organisms are greatly affected and frequently even directly determined by the individual geological processes occurring in the Earth's crust, from the processes of weathering to possibly as far as the movements of the continents caused by continental drift.

favourable for life.[6] And it is because of these stabilizing mechanisms that many of them are willing to consider Gaia to be a single enormous superorganism. In this, they understandably irritate many evolutionary biologists who continue to repeat their magic formula: There is only one Gaia and thus she cannot be subject to natural selection and consequently adaptive traits could not be formed in her. In this case, evolutionary biologists are right and, then again, they are not right. Gaia truly cannot be subject to natural selection and thus the adaptive traits that we know in organisms cannot develop. Nonetheless, Gaia can gradually accumulate properties and mechanisms that contribute to her stability. Similar to any other complex system, Gaia is also subject to the process of **sorting from the standpoint of stability** (see Chapter 5). Both stable and unstable systems are being constantly formed on the Earth. As we mentioned in Chapter 5, there are apparently more of the unstable variety. However, the stable ones, i.e., for example, those containing negative feedback, survive longer. Consequently, Gaia can truly systematically accumulate diverse regulation mechanisms that are capable of ensuring the stability of her environment. Thus, the driving force for the evolution of Gaia is not natural selection, as is automatically frequently assumed by evolutionary biologists and a great many advocates of the Gaia theory, but sorting from the standpoint of stability.

The dove and the hawk or he who plays hasn't time to get into trouble

The fact that biological fitness is not a constant number, but rather a quantity that is highly dependent on the representation of the relevant attribute in the population, has a fundamental effect on the progress of evolutionary processes. Therefore, it has also a fundamental effect on the validity of evolutionary models explaining the development of populations and species on the principle of competition of individuals for the greatest biological fitness. Instead of the former simple laws that could be described using a simple trinomial, the much more sophisticated methodical instruments of game theory are required to describe evolutionary processes.

This brings us back to the evolutionarily stable strategies of J. Maynard Smith and George Price. These British scientists were the first to successfully use game theory to resolve evolutionary problems. It is probable that attempts to use similar approaches appeared sooner; however, they were the first to dress it in a sufficiently attractive cloak of a comprehensible verbal model.[7] Imagine a group of animals of a single species competing for food. If two individuals find the same piece of food, each of them can, in principle, behave in two ways. He can choose the dove

Box 7.4 Game theory

This is an area of applied mathematics that attempts to analyze the progress and results of the competition of individuals that utilize different strategies to maximize their returns. Game theory indicates that the success of individual strategies is frequently dependent on their representation in the population. Some strategies that are very successful when rare can be very unsuccessful when used more frequently.

strategy, i.e. calmly share the food with his competitor, or he can choose the hawk strategy and attempt to drive the competitor away by force or overcome him in battle. If two individuals favouring the dove strategy meet over a piece of food (we will no longer speak of two individuals favouring the dove strategy, but of two doves), they will share the food and each will obtain one half. However, if two hawks meet in a similar situation, they will fight over the food. The winner will obtain all the food and possibly some injuries in the battle and the loser will only carry away his injuries (and his wounded ego). Thus, on an average, hawks will again obtain half the piece of food and half the injuries. An interesting situation occurs when a dove and a hawk meet over a piece of food. The hawk simply drives the dove away and obtains all the food, while the dove leaves, hungry but uninjured.

Imagine that each of these strategies (dove and hawk) is determined by a different variant (allele) of a certain gene. The number of descendants that an individual produces in his lifetime depends on the amount of food that he manages to obtain and the number and extent of the injuries that he suffers in his lifetime. The extent of injuries has a negative effect on the amount of food that an individual finally obtains or the amount of energy gained from food that he can invest in reproduction. If, during their lifetimes, hawks gain, on an average, more of this energy than doves, they will multiply faster on an average and the representation of genes for hawk strategy will increase in the relevant population. If a hawk meets a dove in battle, the hawk will clearly be the winner. Thus, at first glance, it would seem that the hawk strategy is better than the dove strategy and that hawks will come to predominate in the population. However, the reality is different. As soon as hawks begin to predominate substantially in the population, two hawks begin to meet more frequently than a hawk and a dove. However, hawks come out of these battles much worse off than when two doves meet (in addition to half the food, the hawks also suffer half the injuries) and are often worse off than a dove that met a hawk. Half the food need not compensate the hawks for half the injuries so that doves, who don't take away anything from the conflict, are basically better off. Imagine a situation where the population consists of only hawks and one dove enters this population. The dove is defeated in every encounter with a hawk so that it gains no food from these conflicts. Occasionally it encounters a piece of food even when a hawk isn't nearby, so that it doesn't die of hunger. There are so many hawks in the population that, if they meet someone else over a piece of food, it will most probably be another hawk. Thus, it must almost always fight over the food and thus the overall balance of gains and losses from conflicts is negative. A dove with average zero gain is overall better off, can multiply faster and the representation of dove genes will gradually increase in the population. On the other hand, imagine a situation when a single hawk enters a population consisting of only doves. When it encounters someone over a piece of food, it will always be a dove. Thus, the hawk will always win all conflicts (without injuries), will gain the most food of all and will thus be able to rapidly reproduce. Thus, the number of hawks in the population will increase over time. Consequently, which strategy is evolutionarily preferable, the hawk or the dove? Which provides its carriers with the greater biological fitness? Which will

Box 7.5 The Prisoner's Dilemma and the Tit for Tat strategy

The Prisoner's Dilemma game is a favorite subject of analysis for theoretical biologists. This game has many versions, one of which can, for example, be described as follows: Two offenders were caught after they committed a serious crime. There is no direct evidence against them so that, if they cooperate, i.e. deny their guilt, no one can prove their main crime and they will be sentenced only for secondary crimes, such as having possession of a stolen object, with a relatively milder punishment, for example, 3 years in prison. The prisoners are closed in their separate cells and each receives the following offer. If he confesses first and designates his accomplice as the principal guilty party, then he will receive only a mild punishment, for example, one year in prison. However, if he denies his guilt, while the other prisoner who received the same offer, confesses first, then he will receive a sentence of many years. However, if they both betray their accomplices, each will receive a sentence of 5 years in prison. In theoretical studies, the game is played for points rather than years in prison. Usually a game is analyzed in which the reward is 3 points for mutual cooperation, 1 point for mutual betrayal and 5 points for the betrayer and 0 points for the betrayed in one-sided betrayal. Mathematical analysis of the problem demonstrates that, under the given conditions, it is preferable for either of the prisoners to immediately betray his accomplice and not expose himself to the risk of being the second to opt for this approach. The course of a large portion of actual interrogation processes indicates that most offenders do not need to be conversant with the mathematical apparatus of game theory in order to find the only right strategy. Situations that are more or less similar to the prisoner's dilemma are, of course, encountered in nature. An individual sometimes finds himself in a situation where he must choose between betrayal, which can bring great profit or only a small loss, and cooperation, which can bring average profit if the partner also cooperates, but a major loss if the partner betrays him. Under conditions where the two partners will not meet again in the future, or where organisms are involved that cannot recognize or remember their former opponents, both individuals will almost certainly make a choice in accordance with the theory of the "always betray" strategy. A different situation occurs if two individuals play the Prisoner's Dilemma game repeatedly and are capable of remembering the course of the last game. Then the Tit for Tat strategy turns out to be very advantageous. This consists in cooperation in the first game and then, in future games, always repeating the strategy of the other player in the previous game. In nature (and human society), the same opponents frequently meet repeatedly. Consequently, a strategy similar to the Tit for Tat strategy is often employed.[8]

predominate in evolution? Only the latter question can be answered simply and unambiguously. Neither of the two strategies will predominate completely. Over time, an equilibrium will be established in the population with a characteristic, particular, constant ratio of doves and hawks. If we also allow for the existence of "mixed" strategy, then the strategy "behave like a dove in a certain percentage of cases and like a hawk in a certain percentage of cases" will predominate. If chance or external intervention causes a deviation from equilibrium of this ratio of doves and hawks, then it will spontaneously return to the original value.

The ratio of doves and hawks in the population or the frequency with which the carriers of mixed strategy act like doves or hawks depends on the specific values describing the behaviour of the system (for example, on the extent or probability of injuries that hawks usually suffer in more serious battles, the level of reduction in the ability to reproduce amongst injured hawks, the probability that a dove will find food when alone or in the presence of a hawk, etc.). Those evolutionary biologists who prefer to sit in front of a computer, rather than alternately heating and cooling invisible DNA samples in plastic micro-test-tubes or counting the number of finches

with large and small beaks in the Galapagos, immediately set about studying evolutionary games. We soon learned what happens when a different type of strategy appears in the population in addition to doves and hawks, for example retaliators, who act like doves, but only until attacked by a hawk, or bullies that initially act like hawk, but run away when they are faced by a true hawk. Remarkable results were obtained in study of games in which the individual participants met repeatedly and when they remembered how the others behaved last time.

Who would still care about biological fitness!

Of course, this is all very interesting and the knowledge gained by studying this model has provided a valuable insight in many areas of the natural and social sciences. However, from the standpoint of evolutionary biology, it is most interesting that, when the number of carriers of a certain strategy in the population decides on the success of that strategy (the advantageousness of certain properties or certain behaviour), biological fitness (at least as it is understood by most Darwinists and Neodarwinists) loses its fundamental importance for determining the course and result of evolutionary processes. Biological fitness was introduced as a quantity that is directly proportional to the number of descendants that, on an average, the carrier of a certain property (alleles or combinations of alleles determining the particular property) passes on to the next generation. If this number of descendants is dependent on the number of the carriers of the particular property in the population, then this value cannot be determined in any way. For example, we would have to say whether we are interested in the number of descendants in the population otherwise formed by the carriers of the same or the opposite properties or perhaps the same number of carriers of both properties. The parameter of biological fitness, which was originally intended to encompass all the properties determining biological success or lack of success of an individual bearing a particular allele, thus loses its original meaning.

Instead of biological fitness, evolutionary stability becomes decisive for the evolutionary fate of the allele. **Evolutionary stable strategy is a strategy that, once it predominates in evolution, cannot be subsequently replaced by any other strategy.** Evolutionarily stable strategy need in no way be identical with strategy ensuring its carriers the maximum biological fitness.[9] Imagine that a population is formed of only doves. In this case, the population does not lose part of its resources in mutual combats and the average number of progeny corresponding to one individual is certainly higher than in a population that is in equilibrium and that contains both doves and hawks (or only individuals that, with a certain probability (and this probability can be calculated), will behave like doves and with a certain probability like hawks). However, if a mutant or immigrant behaving consistently like a dove appears in a population formed of individuals behaving with a certain probability like doves and with a certain probability like hawks, then it has no chance and its descendants will be eliminated from the population. On the other hand, if a mutant bearing mixed evolutionarily stable strategy "behave with a certain probability like a dove and with a certain probability like a hawk" appears in a population formed only of doves, then it will gradually predominate even if the average biological fitness of the members of the given population decreases compared to the initial state. **Darwin's concept (repeated in all textbooks of evolutionary biology to the present day) that individuals with the largest biological fitness value will predominate in evolution is, at the very least, misleading (and, if fitness is understood in the usual Neodarwinist manner, in fact erroneous).**

Thousands of hawks and doves in us

How common is frequency-dependent selection in nature? Hard to say. However, I dare to guess that the biological fitness of almost any trait depends at least partially on its frequency in the population. In a great many cases, this dependence is positive, i.e. as the frequency of the trait increases in the population, the fitness of its carrier increases. In this case, there is fundamentally no difference between biological fitness and evolutionary stability.

When considering the commonness of frequency-dependent selection, it is primarily necessary to be aware that the evolutionary game is not related only to genes affecting the behavior of their carriers. The subject of frequency-dependent selection can be practically any trait from the colour of fur to the activity of any cellular enzyme. Cases where a predator selects the more common form of prey are very common in nature. A situation often occurs where two forms of a particular species of fauna or flora are specialized in utilizing two different natural resources where, at a particular instant, the more successful and thus more numerous form rapidly uses up its resources and is thus penalized by natural selection in the subsequent generations. In sexually multiplying species, which are in the vast majority in nature, we can find one more quite fundamental reason for the frequency dependency of biological fitness, related to practically all genes and all traits. **The manner in which the presence of certain gene variants (i.e. alleles) affects a particular trait and thus the overall appearance and behaviour of the individual, is dependent, amongst other things, on which variant of the given gene is located on the second copy of the relevant chromosome that the particular individual inherited from the other parent.**

If we live in equatorial Africa and have inherited from our mother a gene (more precisely, the allele of a gene) for a certain uncommon form of the red blood protein haemoglobulin, we could have much greater resistance to malaria and thus substantially greater biological fitness than our contemporaries who did not inherit the allele. However, if the particular allele, called the S allele, occurs in the local population with high frequency, there is a substantial chance that we will also inherit the same allele coding unusual haemoglobulin from our father. In this case, we would be very unlucky, and there would be no increased biological fitness, in fact there would be no biological fitness at all, as we would die of a serious form of sickle cell anemia before we would manage to reproduce. Similarly as in the above-described hypothetical example of a dove and hawk meeting over a piece of food, in this and a great many other cases, a pair of alleles of different genes meet after the combination of the male and female sex cells in the newly formed zygotes, from which a new individual is, or is not, formed.

Box 7.6 Why aren't there two-headed mutants running around after Chernobyl?

Because the relevant embryos with developmental defects were not embedded in the uterus and were aborted. This process has been studied in detail in rodents and in cattle exposed to high radiation levels in the vicinity of the Chernobyl nuclear power plant. These animals did not produce more deformed young, but their fertility was substantially reduced. And that is not all. It was found that, under the new conditions of elevated radiation, heterozygote individuals very frequently produced, with different probability, descendants with one or the other allele – i.e. exactly that phenomenon that we discussed in connection with the S allele. By the way, infant mortality increased substantially in Eastern Europe after the Chernobyl nuclear power plant exploded. It can be expected that a similar change (in this case, on the other hand, a decrease) could be found in the fertility of the inhabitants of Europe (e.g. in the average period of time that a pair waited for a child). It is quite probable that, in addition to reducing the population growth, the explosion of the Chernobyl power plant was also manifested in the composition of the gene pool of the European population (because of the above-mentioned changes in the probability of transmission of the individual alleles to the next generation).[10]

For incorrigible cavillers, I would like to add: the fact that a new individual need not be formed from the zygote is of quite substantial importance. If a dove and a hawk meet over a piece of food, each takes away a different reward from the conflict. On the other hand, if the individual is a heterozygote, i.e. if normal alleles and S alleles met in his cells, it might seem that they both take the same reward from the meeting, as his biological fitness is affected to the same degree by the evolutionary fate of both alleles. In actual fact, this need not be true. Imagine, for example, that a large percentage of the zygotes carrying two S alleles do not settle in the sex organs of the woman and disappear without substantially utilizing the resources of the maternal organism – i.e. if we neglect the fact that the woman does not become pregnant in that month.

Side track: I do not have a sharply defined opinion in respect to artificial interruption of pregnancy, but don't find either of the extreme positions attractive. I would certainly be interested in discovering whether the proponents of complete prohibition of artificial interruption of pregnancy are at all aware that the vast majority of pregnancies are terminated by spontaneous abortion during the first few days or weeks after formation of the embryo, i.e. long before the woman even discovers that she is pregnant. And this is a very good thing, because most of these aborted embryos carry genetic defects. I am quite shaken by the idea that a proponent of the rights of the embryo might one day get the brilliant idea that it would be possible (and a good thing) to employ some sort of pharmacological intervention to turn off the mechanism ensuring control of embryo quality. End of the side track and back to the original problem.

If embryos with two S alleles are rarely implanted in the uterus or embryos with two S alleles are frequently aborted, each of the two alleles will leave the meeting of normal and S alleles in the genome of a heterozygote woman with a different result. If the partner of the woman is also a heterozygote, then the number of descendants bearing a copy of the normal alleles of the mother will be greater in the final analysis than the number of descendants bearing copies of her S alleles.

Box 7.7 The effect of the parasite *Toxoplasma gondii* on the sex ratio of human beings

As mentioned in Chapter 2, many people are infected by the parasite *Toxoplasma gondii* throughout their lives. In developed countries, 15–40% of women of reproductive age are usually infected; in developing countries with lower hygienic standards, the occurrence of this "latent" toxoplasmosis approaches 90%. A study that we recently performed on a large number of women indicated that far more boys than girls are born to infected women in their first pregnancy. In a set of 111 women with the highest antibody levels (i.e. with the strongest or freshest, nonetheless latent, infection), the ratio of boys to girls attained a value of 2.6:1. We observed a similar phenomenon in experiments performed on infected mice.[11] It is not yet known in which way toxoplasma affects the ratio of the sexes of human beings. However, it seems most probable that the protozoa in some way reduces the probability that the embryos of individuals of male sex will be aborted in the first weeks of pregnancy. It is well known that male embryos have a much better chance of implantation in the uterus of the mother than female embryos, but that they simultaneously have a much greater chance that they will be aborted in the first weeks of pregnancy. Of a ratio of the sexes of 1.64:1 in favour of boys in the 5–7th week of pregnancy, the secondary sex index usually decreases by the time of parturition to the usual value of 1.06:1, corresponding to 106 newborn boys to 100 newborn girls. The immune system of the mother plays an important role in elimination of male embryos as it recognizes antigens specific for male cells, H–Y antigens. It is known that toxoplasma has a substantial impact on immune processes occurring in the infected organism. Thus, it is possible that toxoplasma can affect the ratio of the sexes in favour of boys by suppressing the component of the immune system that is responsible for elimination of male embryos. In conclusion, two questions to make you think: Older parasitologists observed that, in a population in which about 30% of the individuals are infected by toxoplasma, more than 80% of children with Down's syndrome are born to mothers infected by latent toxoplasmosis. Modern parasitologists and physicians, of course, laughed at this – we obviously know that Down's syndrome is not caused by a parasitic protozoa but by the fact that two copies of chromosome number 21 accidentally entered the egg during meiosis and the individual was created by the fertilization of this egg by normal sperm so that the individual has three copies of this chromosome in their cells instead of two. The first question – how could toxoplasma lead to increased frequency of children with Down's syndrome in infected mothers, without having to attack the future sex cells and play around with their chromosomes during meiosis? Second question, far more difficult, to which I also do not know the answer – should parents who are taking care of a beloved child with Down's syndrome curse toxoplasma or thank it?[12]

This will be because the embryos with the S alleles of the mother, which also carry the S alleles of the father, will most probably be aborted. However, if the partner is a homozygote with two normal alleles, the mother will transfer to her descendants the same number of copies of both her alleles. On the other hand, in this case, descendants bearing copies of normal alleles will frequently die in childhood of malaria, while the descendants bearing copies of the S alleles will be far more likely to live to reproductive age. The rules of the relevant evolutionary game are thus, in actual fact, far more complicated than in the case of the model of the dove and the hawk. And we are completely ignoring the fact that not only the alleles of a single gene, but also the alleles of various genes can interact in their effects. Thus, the biological fitness of the carriers of certain alleles very frequently depends on which alleles are present in the other parts of the genome. We will return to this important phenomenon in the next chapter but one.

The revolution that didn't happen

How is it possible that the discovery of evolutionarily stable strategy in the 1970s did not completely wipe out the Neodarwinist theory of adaptive traits based on selection of individuals with a high biological fitness value? How is it possible that, even in the most modern textbooks of evolutionary biology (with rare exceptions[13]), the aspect of evolutionarily stable strategy is mostly encountered only in chapters devoted to evolutionary behavior or evolutionary altruism? It's hard to say. Textbooks mostly multiply by cloning – more or less creative reworking of older textbooks. Scientific workers nowadays mostly do not have time to excessively mull over the theoretical basis of their field. We have to write grant proposals and grant reports and, if we ever think about paradigms forming the basis for our work, this is usually so that we can manage to fit our results into their framework.

And the alert guardians of the cathedral of science (reviewers and editors of professional journals) deal with someone who doesn't manage to fit his results into their framework according to the ancient and well-tried rules for dealing with heretics. Some tried-and-true procedures are not employed in the present-day generally soft times. An original, but careless author thus does not generally get burned at the stake but, nonetheless, is faced by a fate far worse than death, i.e. returning of his manuscript from all editorial boards and thus general oblivion.

Box 7.8 Grants and grant reports

A large proportion of funds for science are obtained by scientists, not from their employers, a research institution or university, but rather from specialized national or international grant agencies. A scientist thinks up an interesting and feasible grant project, describes it in detail and exactly calculates the funds required to resolve it. Then he or she submits the proposal in a grant competition, announced by the individual agencies, usually once annually. The officials first exclude all the proposals that did not meet the relevant formal requirements (form B-6 was not accompanied by a confirmation from the Vice-Dean for public relations that the animal facilities do not currently keep duck-billed platypuses infected either with foot and mouth disease or bird flu) and then send them to a number of scientists (usually applicants from previous years – whose addresses they have in their databases) for expert evaluation. On the basis of the expert reports obtained, a commission of experts of the particular grant agency (consisting of scientific workers who keep an eye on one another) establishes an order of the submitted proposals and a few percent of the best projects are then financed. Projects usually last three years and each year the responsible worker submits a report on the results obtained and the manner of managing the funds. The present system has the great advantage that it tends to favour the capable and hard-working rather than the incapable and lazy, that it promotes cooperation amongst the employees of a single institution (who are not competing for a joint package of institutional funds) and that it limits the potential for intervention by easily corruptible officials and politicians. It has the disadvantage that it tends to favour short-term projects with predictable outputs, that chance plays a considerable role in the allocation or non-allocation of funds, specifically in the reviewers that receive the project for evaluation and their momentary moods, and also that creative scientific workers are buried under mountains of paperwork. It is said, but this will most probably be only a rumour, that experienced scientists write grant proposals for projects that they already have more than 75% completed. They then use the allocated funds for work on new projects that, if they turn out well, can become the subject of the next grant application. And worst – some of us even insist that there is no other reasonable approach as it is a well known fact that scientific discoveries can, in actual fact, not be planned in advance.

The discoverer of the theory of evolutionarily stable strategies John Maynard Smith was for many years the most respected representative of European evolutionary biology.[14] He wrote a number of excellent textbooks and monographs concerned with the various aspects of evolution. If I am any judge, he was not only an original thinker with a deep understanding of the subject of evolutionary biology, but also a fascinating personage, who had a very good chance of introducing quite revolutionary changes into the evolutionary paradigm. I find it hard to believe that Sir John was not aware of the ramifications of his discovery. It almost seems more probable that he was very well aware of this, but did not want to deprive us younger and less capable scientists of the pleasure of making these discoveries ourselves. I am also fond of the fantasy that he is now looking down on our ridiculous bustling around and is quietly telling us – completely cold, not even getting close yet, getting warmer, …

Summary and incitement

To summarize. Situations occur very frequently in nature in which the biological fitness of the carriers of certain alleles depends on the frequency of these alleles in the particular population. This means that the overall chance of certain alleles in evolution cannot be expressed in terms of a certain number expressing the relevant selection coefficient or biological fitness, as has been done by generations of evolutionary biologists, but that it must be described by a more or less complicated function. Consequently, the progress of competition of various alleles cannot be described by the laws of selection derived in the past by geneticists, but the mathematical apparatus of game theory must be used for this purpose. The fate of the individual alleles is not decided by how each of them affects the average biological fitness of members of the population, but by which of them determines an evolutionarily stable strategy, i.e. a strategy that, when it once predominates in the population, cannot be replaced by any other strategy. This discovery, which was made in the 1970s by John Maynard Smith and George Price, actually constitutes a quite fundamental blow to the very foundations of the Darwinist theory of the evolution of adaptive traits. The next chapter describes the theory of the selfish gene which basically constituted a fundamental deviation from Darwinist theory. It is probably not in good form to shout about it too much, but the theory of the selfish gene shows that, in sexually reproducing organisms, evolution of adaptive traits occurs through a quite different mechanism than that proposed in the middle of the 19th century by Charles Darwin.

Footnotes

1. The theory of neutral evolution is described in the works of Moto Kimura – e.g., the review article published in the *Japan Journal of Genetics* 68: 521–528, 1993, or the monograph *The Neutral Theory of Molecular Evolution*. Cambridge University Press, Cambridge (1983).
2. An experiment with guppies was published by L.A. Dugatkin in L.A. *Proceedings of the National Academy of Science, U.S.A*, 93: 2770–2773, 1996.
3. However, nature is diverse. In some species, in con-

trast, we encounter a situation in which females prefer males with unusual phenotypes, see, e.g., *Current Science* 78: 141–150, 2000. Surprisingly, the two phenomena are not mutually exclusive and can be active in a single population simultaneously. If an unusual male appears in the population, he will tend to have below-average success amongst most females. However, he could be "just the thing" for some females and, as a consequence, will reproduce (although only with certain females) more readily than the average male in the population.

4. I borrowed this nice example of frequency-dependent selection in orchids (of course without Hansel and Gretel) from the textbook by Freeman S. and Herron J.C. *Evolutionary Analysis*. Pearson Education, Inc. (Upper Saddle River, N.J.), 2004. Originally, the relevant study was published in the *Proceedings of the National Academy of Science, U.S.A.* 98: 6253–6255, 2001.
5. A study documenting the connection between the frequency of left-handed people (3–27 %) and the number of murders in different traditional societies was published in *Proceedings of Biological Sciences* 272: 25–28, 2005.
6. All of the books of J.E. Lovelock are definitely worth reading. You could begin with J.E. Lovelock, *Gaia: A New Look at Life on Earth*, Oxford University Press, 1979.
7. I would be quite interested to discover if this model would have had the same success if deeply believing Price had had his way at the time and introduced for his model the name Mouse and Hawk to avoid undesirable religious associations. The original article can be found in *Nature* 246: 15–18, 1973.
8. The Tit for Tat strategy was described in *Science* 211: 1390–1396, 1981 and other strategies are described in *Nature* 355: 250–253, 1992, *Nature* 364: 56–58, 1993, *Journal of Ethology* 19: 1–8, 2001.
9. I don't want to hide the fact that this problem is much more complicated than can be described in this context. For example, in a structured population where individual local populations are constantly established and disappear, the criterion of success can be not evolutionary stability (which is basically the ability of the strategy to manage the best of all the possible strategies in competition with itself) but rather evolutionary invasiveness – the ability to compete with other strategies (predominating in the population). It is possible that most species live in such structured populations and that evolutionary invasiveness is the most important criterion of evolutionary success for them. However, I think that the main message of the chapter is valid – biological fitness, as a constant assignable to a certain mutation, both does not exist and is certainly not a measure of future evolutionary success.
10. I can highly recommend the article *Genetic Consequences of Chernobyl* (in Czech) published in: *Vesmír* 85: 201–208, 2006.
11. Our article on the fact that toxoplasmosis leads to an up to 2.6-fold increase in the probability of birth of a descendant of the male sex in humans was published in *Naturwissenschaften,* 94: 122–127, 2007, after being rejected by eight other journals. In four cases, no review was performed; I cite the response from the journal *Nature* (and thus probably eliminate the possibility of publishing anything in this journal in the future): "We do not doubt the technical quality of your work, or that it will be of interest to others working in this and related areas of research. However, we feel that your findings are of insufficiently immediate interest to researchers in a broad range of other disciplines to justify publication in Nature.". It was accepted in the last journal only because the editor-in-chief noticed that its results are immediately related to those of another article that was undergoing the review process and, after receiving the opinion of the subsequently addressed statistician, she decided to overlook the more-or-less negative recommendations of the reviewers. The manuscript of an article on the probability of birth of males in mice was accepted to the fifth journal (*Parasitology* 134: 1709-1718, 2007) and an article on the protective effect of the Rh factor has been sent to twelve journals (only two editors sent the manuscript to referees); this article still has a long way to go.
12. This extremely interesting result, indicating that *Toxoplasma* apparently turns off or weakens some processes of embryo control, was observed in a study by the founder of modern Czech parasitology, O. Jírovec, *Československá pediatrie*, 12: 713–723, 1957. The results of this work are very convincing. Of 94 mothers of children with Down's syndrome, 84% were infected with *Toxoplasma*, while only 32% of women of the same age in the normal population were infected (of 38 fathers of children with Down's syndrome, only 24% were infected). The results also indicate that this could be an adaptive property (evolutionary adaptation) of *Toxoplasma* that would allow it to be transferred from the mother to the child. In 50% of cases, children with Down's syndrome with infected mothers were also infected (only 13% of healthy children of the same age were infected), while only one was infected amongst 15 children with Down's syndrome whose mothers were not infected. The elevated occur-

rence of toxoplasmosis amongst patients with Down's syndrome was also described in the work of British authors in the *Journal of Hygiene Cambridge*, 63: 89–98, 1963 and in a number of other studies.

13. With my typical self-effacement, I would like to point out that one of these rare exceptions (and I know of no other) consists of my textbook on evolutionary biology.
14. Practically all of the books of Maynard Smith can be recommended. I especially recommend Maynard Smith, J. & Szathmary, E. (1995), *The Major Transitions in Evolution*. W.H. Freeman Spektrum, Oxford (personally, I would recommend this original version of the book, rather than the later, abbreviated version intended for the broader public, modified according to the ideas of the publisher) and also books which Maynard Smith wrote as the only author, *The Evolution of Sex*, Cambridge University Press, Cambridge (1978), *The Theory of Evolution*, Cambridge University Press, Cambridge (1993), *Evolutionary Genetics*. Oxford University Press, Oxford (1998). These books are not intended primarily for the lay public; however, if you skip the text containing mathematical formulae (which 90% of readers do anyway), they are quite readable.

Chapter 8 Formation of the theory of the selfish gene – Darwin, watch out – someone is after your throat

While the theory of evolutionarily stable strategy didn't cause a great disturbance even amongst the professional public, the theory of interallelic selection of William D. Hamilton attracted well-deserved attention from both the professional and lay public.[1] It had the good luck that it was soon taken over by another British evolutionary biologist, the excellent proponent of evolution Richard Dawkins, who sewed an attractive, sexy coat for it, which is now mostly called the **selfish gene theory**.[2]

The selfish gene – finally, something is happening

Only after extraction from the language of mathematical symbols into a "user friendly" form of description of actual biological objects was it found that the theory of interallelic selection or, rather, the new theory of the selfish gene moves the contemporary model of evolution outside of the area of Neodarwinism. When we read the first edition of the "Selfish Gene" and the postscript to the second edition, we find to our surprise that the actual importance of the new theory became apparent to the author of the book himself long after he completed his text. In the first edition, he only stated that the new theory is only a different, although apparently better way of looking at Darwin's old theory. He illustrated his opinion on the example of the Necker cube, on a normal two-dimensional depiction of a cube that cannot be differentiated as to whether it corresponds to a three-dimensional cube depicted from the top or the bottom. Both of the ways of understanding the picture are equally good, but they are mutually incompatible and our brain is capable of flipping back and forth between the two views of the system of lines on the flat paper; try this yourself on Fig. 8.1. It was only after the next edition of the "Selfish Gene" that Dawkins permitted himself to state that the classical and new views of biological evolution are not equivalent, that they lead to different conclusions in certain situations and thus that only one of them can be correct. It is quite possible that Dawkins came to this conclusion much sooner and that he attempted to camouflage this in the first edition of his book. It is not very tactical to announce publicly that our theory is different and better than the theory that everyone else has held until now. Science is very democratic in a certain sense. Basically, a kind of vote is taken on which theory is better. If the majority of scientists conclude that they find some theories or some authors unattractive, then they simply won't mention them in their works and the particular theories and authors fall into oblivion. On the other hand, if a particular theory or its author is attractive for most research workers, then they will frequently cite this theory and, on the whole, it is not important whether

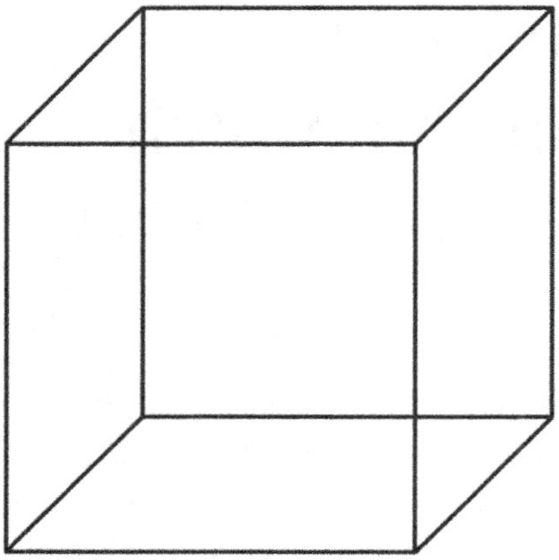

Fig. 8.1 The Necker cube. If you look at the drawing of the cube for a while, your brain will switch back and forth between two different perceptions of the figure – between the cube depicted from above and from below.

the theory is actually correct. Anyone who behaves differently, for example, cites an unpopular author or, heaven forbid, fails to cite a popular author, then he risks that he will also become an unpopular author. Dawkins had already gotten himself into enough trouble by addressing his book to the general public instead of honestly and properly publishing his opinions in specialized journals that are inaccessible and would, anyway, be unintelligible for nonprofessionals and, for professionals, also almost (financially) inaccessible and almost unread (for lack of time).

And to explain the fact that professional articles do not get read: with the present system of managing science, we don't read other people's articles: we have to save time somewhere so that we can write our own articles. The approach proposed by colleague Zrzavý, that we read only articles in which our own name appears (I hope that I understood him correctly and that he didn't mean articles in which his name appears) so far seems to me to be too radical.

Authors who are capable of explicating on the knowledge and theory of their field in a form that is comprehensible to the general public are not greatly respected by their colleagues. It is not important whether they also produce their own good-quality scientific results or not – if someone simultaneously writes successful popular scientific books, it is immediately obvious to everyone

Box 8.1 The financial inaccessibility and unread nature of professional journals

The publishing of most professional journals has been taken over by commercial publishing houses and their prices have been cranked up to absolutely impossible heights. It has been calculated that, while publication costs for one article work out at approximately $ 500, libraries around the world pay a total of $ 16 000 for one article (costs estimates in 1999). This sum is comparable with the average costs for the research itself, which correspond to approximately $ 20 000 per published article. The greater part of this sum ends up in the pockets of commercial publishers of professional literature. A few years ago, scientists attempted to rise up against the dictate of commercial publishers and began to massively sign a petition exhorting other scientists to boycott journals that do not publish an electronic version of articles that is freely accessible to the public on the Internet within six months of publishing the printed version. Of course, the boycott was unsuccessful and most authors continued to send their manuscripts to the relevant journals. When someone organizes the next boycott, I would like to suggest a "minor" adjustment – so that the boycott doesn't hurt the boycotter more than the boycotted, it should not consist in not sending manuscripts to expensive journals that do not publish an electronic version, but in not citing works "published" in these journals. The decrease in the impact factor of the journal will certainly make the publisher see sense very quickly.

Box 8.2 Sexual and asexual organisms

To a first approximation, the situation is clear. Some species of organisms reproduce sexually, i.e. their descendants are formed by the merging of the sex cells of two organisms. Others reproduce asexually, i.e. their descendants are formed by splitting off of part of the parent organism (e.g. tubers for potatoes) or from individual specialized cells intended for this purpose (e.g. some species of stick insects and fish). On closer inspection, the location of the boundary between sexual and asexual reproduction is less clear and opinions of professionals on the difference between sexual and asexual reproduction need not completely agree. However, in this book, I will stick to the approach that considers asexual reproduction to be the formation of descendants with a genotype identical with that of one parent and sexual reproduction to be the formation of descendants with a genotype formed by the combination of the alleles of two parents. The fact that sexuality was almost certainly not originally connected with reproduction and fulfilled a completely different function is something I prefer to leave out here – this could even be the subject of a separate book.

(i.e. everyone that is not capable of this – and that is most of us) that he must be a complete professional imbecile.[3]

Now, what is the selfish gene theory about? The classical Darwinist (to be more exact Neodarwinist) model of evolution of adaptive traits through the action of natural selection can function very well for organisms without sexual reproduction, for example amongst bacteria, but it cannot function amongst organisms reproducing sexually.

When mutant bacteria divide asexually into two daughter bacteria, they transfer to each of them one identical copy of their genome. As a consequence, their descendants also inherit the properties determined by the particular mutation. If the mutation increased the viability or fertility of its carrier, then these properties are also manifested in its descendants and these descendants will multiply faster in comparison with the other members of the population. Thus, the number of carriers of the mutation will gradually increase in the population until they finally completely replace the original unmutated form. The process of fixation of mutations will be continually repeated in evolution, so that the bacteria will gradually acquire new adaptive traits and will become increasingly adapted to their environment.

This is classical Darwinism. However, a problem occurs in sexually reproducing organisms in relation to inheritance of advantageous (and disadvantageous) traits. They do not reproduce by copying their genome and transferring a copy to the next generation. The basis for their reproduction consists in the formation of sex cells containing only one half of the genetic information of the parents and subsequent merging of two gametes and combination of their genetic information, originally derived from two parent individuals. The combination of alleles (variants of individual genes) in the genome of an individual (i.e. the genotype of the individual) fundamentally affects the traits of the individual, their appearance and behaviour, i.e. their phenotype. In case of asexually reproducing organisms, the genotype of the individual is inherited (maximally "enriched" by an occasional new mutation). Because the genotype of an individual to a substantial degree also determines his phenotype, the phenotype of the individual is also logically inherited. (Of course, the phenotype is also affected by the effects of the external environment, such as nutrition.)

However, the genotype is not inherited in sexually reproducing organisms; rather it is formed anew in each individual from newly mixed genes, more exactly alleles, derived equally from both parents. Consequently, the phenotype of the individual and their biological fitness are also not inherited. The biological fitness of the individual is not determined so much by individual advantageous mutations, but far more by the particular fortunate or unfortunate combination of alleles that together form an advantageous or disadvantageous genotype, and thus a more of less functional phenotype. In the first generation, a child has a chance to inherit at least a similar combination of alleles to that of his parents. However, this probability is very rapidly reduced in subsequent generations. If you had biologically very good quality grandparents, you have a very low chance of inheriting their quality to a greater degree than any other of your contemporaries. From your great grandparent, you can inherit his name (if this is a predecessor in the direct paternal line), his house, enemies and debts. However, his unique genotype has long been scattered in the overall gene pool of the population. The probability that it will emerge again in a similar form in some other member of the population is very small and can occur with equal probability amongst his direct descendants and amongst the descendants of any other member of the population. (This is, of course, under the more or less reasonable assumption that you don't reproduce by cloning – this is forbidden; if you do it, you should be ashamed – or that you are not a member of an ancient Egyptian dynasty and do not reproduce over a great many generations exclusively by the brother-sister system.)

However, if biological fitness is not inherited amongst sexually reproducing organisms, then the basic mechanism and main driving force for Darwinian evolution, natural selection, can also not function. Thus, Darwin's model of biological evolution is capable of explaining the development and formation of adaptive traits amongst asexual bacteria, but cannot satisfactorily explain the development and formation of adaptive traits in species reproducing sexually, i.e. in the vast majority of species occurring on the planet, from the liver fluke to the chimpanzee.[4] Nonetheless, we are surrounded by a thousand and one examples of the fact that biological evolution occurs in both asexually and sexually reproducing species. In addition, palaeontological records and the results of comparative biology have shown that evolution occurs much faster amongst sexually reproducing species than amongst asexual species. Fossil bacteria from rocks more than 3 billion years old can mostly be readily assigned to contemporary genera. A great many people believe that it was the evolutionary discovery of sex that started the evolution of modern eukaryotic organisms sometime before one and a half billion years ago.

How can the theoretically derived impossibility of functioning of Darwinian evolution be brought into accordance with the extensive empirical evidence confirming its existence, specifically with the occurrence of adaptive traits in all types of organisms? In his book, "The Selfish Gene", Dawkins tried to show that this is not so difficult. Basically, it is sufficient to abandon

Box 8.3 Eukaryotic and prokaryotic organisms

The original organisms occurring on the Earth were prokaryotic, i.e. their cells did not have a classical nucleus and a number of other organelles. Of contemporary organisms, two groups, bacteria and the less well known archaea, are prokaryotes. Eukaryotes developed much later from prokaryotes. They have much larger cells, in typical cases their volume is greater by 3–4 orders of magnitude; they contain a nuclear cell wrapped in a double membrane and a number of specialized organelles, of which some, specifically mitochondria and chloroplasts, were formed at some time in the past by "taming" prokaryotic organisms – possibly parasitic bacteria related to present-day rickettsia (mitochondria) and algae (chloroplasts). A eukaryotic cell is actually a sort of conglomerate (chimera), formed in the past by the combination of several prokaryotic organisms belonging to both the group of archaea and the group of bacteria. For this reason, mitochondria and chloroplasts continue to bear their own genome – residues of the DNA of the original symbiont. In the framework of eukaryotic organisms, multicellular organisms developed over time, for example plants and animals. Prokaryotic organisms remained single-celled, but frequently form colonies of cooperating cells (belonging to a single species or to several species).

the traditional Neodarwinist view of biological evolution as the competition of individuals for resources and the fastest possible reproduction and, in its place, to concentrate on the level of the genome itself. At this level, the conditions for the action of classical Darwinist evolution are much more favourable. The individual alleles reproduce by copying and are consequently transferred from one generation to the next in unaltered form. Obviously, an allele is not an indivisible bead, but a chain of nucleotides that can be cut in half at any time and rejoined in the process of genetic recombination. Thus, a third allele can be formed of two alleles by recombination. However, on the other hand, this is such a short chain that the probability that a specific gene would be altered from one generation to the next by recombination is very small. It is known that, on an average, one recombination event (crossing over) per chromosome occurs in the formation of a sex cell. Thus, if our great grandparent had a certain rare variant of a gene in his genome and if he had a sufficient number of great grandchildren, then it is almost certain that a number of them will carry the same variant of the particular gene in their genomes as he did.

Puppet theatre

How can this be reconciled with the model of biological evolution based on natural selection? Simply. In evolution, in actual fact, the individuals of a certain species do not compete as to who will leave the greatest number of children behind them (as assumed by Darwin), but the various alleles of the particular gene compete as to which will be transferred in the greatest number of copies to the next generation (as shown by Dawkins). Alleles use various tricks for this purpose. For example, some of them are capable of forcing the cellular apparatus that was originally intended for repair of damaged DNA to modify (rewrite) the variant of the gene located in the same position of the chromosome derived from the other parent according to its own sequence. Others are capable of forcing the apparatus dividing the nucleus so that, during the formation of haploid female gametes (eggs) from the original diploid cell, the chromosome derived from the

Box 8.4 Is it possible to write in a biological text that alleles learned to do something in order to achieve something else?

Of course, but only in the case that you are not a student who is writing a thesis work and expects to encounter an unfriendly or especially dense reviewer. It is usually completely clear from the context that you are not suggesting that alleles have the ability to plan their future behaviour in relation to achieving some future goal. It is, of course, possible to state quite correctly that plants form attractive flowers because the members of this species who accidentally gained this ability by mutation in the past better attracted pollinators than their competitors and thus had a greater number of progeny, who inherited this ability. However, it will be far easier to understand if we simply state that plants form flowers in order to attract pollinators.

other parent (with a different allele) is sent out to the polar body, where it is destined for destruction, see Box 6.5 on p. 89, so that the chromosome on which it is itself located is preferentially placed in the nucleus of the egg. Others are capable of giving a "kiss of death" to genes on the other chromosome, so that all gametes bearing these "kissed" genes are unviable (e.g. because the relevant sperm will lack tails). Male mice with one normal and one such "nasty" allele (denoted here as the t-allele) will thus produce only sperm bearing copies of the nasty allele and will thus be capable of passing down only this allele to their descendants. However, in evolution, a large proportion of alleles "chose" an apparently simpler, but actually much more cunning strategy. They learned to affect the properties of the individuals in which the genome is located in such a manner that this individual has the greatest biological fitness – in the conditions under which it exists, it produces the greatest number of viable progeny.

Consequently, Darwin and, after him, all the proponents of Darwinism and Neodarwinism thought that individuals competed amongst one another for the greatest biological fitness. In actual fact, this is on the background of the battle of the individual variants of a single gene to decide how many copies of it will be transferred to the next generation. The main battle between the alleles occurs in the "sport" of programming of the properties of the organism. The individual alleles of a single gene compete to determine which of them, through its effect on the growth and development of an individual or in an additional manner on its properties, including behaviour, will create an organism with greater biological fitness, i.e. with better ability to gain resources from the environment, resist enemies, seek out good sexual partners and reproduce. Dawkins showed that all the bustling about of living organisms in the world is just a sort of puppet show, in which the living organisms are only passive actors. Not that the genes would continuously pull strings and thus predetermine each movement of their puppets – the individual organisms. However, genes determined the rules of the game and even created the puppets. But they do not intervene much in the actual course of the play (in the behaviour of the puppets during the "theatre performance") and leave this to the competence of the organs and organ systems of the puppets that are intended for this purpose – for animals, mostly in the competence of their nervous systems. The individual organisms, in Dawkins' terminology vehicles, thus function in nature as completely autonomous robots that their creators (i.e. the genes located in the chromosomes of the cells of the robot) basically endowed with free will or, to be more exact, with the ability to behave in the way that their controlling nervous system (formed by the action of their genes and of their surroundings) dictates.

Fairy tale about evil Blue Beard

At first glance it may seem that there is fundamentally no difference between the classical organism-centred and the new gene-centred views of biological evolution. In both cases, the greatest amount of the competition leading to the development of organisms and the creation and improvement of adaptive traits occurs in the form of competitions among individuals in the population for the fastest reproduction. However, in actual fact, the two models of evolution are not equivalent. **A situation can be found in which the two models yield different predictions related to the future development of a particular system.** The spiritual father of the gene-centered view of biological evolution, William Hamilton (1936–2000) gave a very nice, although only hypothetical, example of such a situation in his original article.[5] As neither he nor, later, Richard Dawkins, named this model, I have taken the liberty of remedying this only serious inadequacy in an otherwise very nice and illustrative model. And, because I am sometimes rather malicious, I chose the name **Blue Beard Model** some years ago, which students of evolutionary biology will be able to easily confuse with the name of a different evolutionary model, the **Green Beard Model**, which Dawkins created and named in his book, "The Selfish Gene" (see the box).

So what does the **Blue Beard model** look like? Imagine a carnivorous animal that lives in pairs and has an average of 8 young, 4 females and 4 males. However, under normal conditions, it is

Box 8.5 The Green Beard Model

In this model, Dawkins shows that the alleles of genes are quite selfish; that each of them is interested only in the number of copies of itself that it can pass on to the next generation and not the number of copies of other genes in the genome, of which it is a part, that are passed on to the next generation. Let's imagine Dawkins' hypothetical green-beard allele, which leads to the formation of a green beard in its carriers and also leads them to assist other "green-beards". It can be seen (and it's very easy to demonstrate on a mathematical model) that such a green-beard allele has a much greater chance of spreading in evolution than an allele that would lead its carriers to help their blood relatives. The carriers of green-beard alleles will pass (and will help to pass) on to the next generation more copies of themselves than copies of other alleles. On the other hand, an allele that would lead its carriers to assist blood relatives will be worse off, even though it would objectively ensure that a greater percentage of all the alleles (of all the genes) of its carrier are passed on to the next generation. However, it would not ensure that carriers of itself would be amongst them more frequently than the carriers of alternative alleles occurring on the second copy of the same chromosome. The unrelenting laws of biological evolution thus mean that each allele will behave quite selfishly and will be completely indifferent to the fates of the other alleles on the same chromosome or in the same genome. The only thing that it will count will be the number of copies of itself that it passes on to the next generation.

Box 8.6 Y Chromosome

In many animals and some plants, the sex of an individual is determined at the time of merging of the male and female sex cells by the presence of sex chromosomes. Mammals are an example of organisms in which the male carries two kinds of sex chromosomes, the X and the Y chromosome, while the female has both sex chromosomes the same, i.e. has two X chromosomes. During meiosis, the two sex chromosomes form a pair and then separate (similar to the other pairs of homologous chromosomes), each to its newly forming sex cell. Thus, two types of gametes are formed in males, one with an X sex chromosome and the other with a Y sex chromosome. Females form only one type of gamete – all with an X chromosome. If, during fertilization, a male gamete bearing an X chromosome merges with a female gamete bearing an X chromosome, then a female embryo is formed. However, if a male gamete bearing a Y chromosome merges with a female gamete bearing an X chromosome, then a male embryo, XY, is formed. The Y chromosome carries genes that are necessary for the formation of male sex organs and the products of these male sex organs subsequently affect the entire development of the embryo in a way such that a male is formed.

capable of feeding only 6 offspring, so that the two weakest usually die when young. Now imagine that mutation occurs on the sex Y chromosome of the male, leading to a new variant of the gene affecting the parental behaviour of males.

The carriers of this allele, i.e. always only a male, as females do not have Y chromosomes, kill all their freshly born daughters and use their meat to feed their sons. What will be the fate of new alleles for this blue beard behaviour? Do they have any hope of spreading in the population or will their bearers be removed from the population by natural selection? From the standpoint of classical Darwinism, we would tend to expect the second alternative. While a normal male would bring to adulthood an average of 6 young from each litter, a Blue Beard would bring up only 4 young. Thus, a Blue Beard has about one-third lower biological fitness and natural selection should place him and all his male offspring (there will be no female offspring) at a disadvantage. However, from the standpoint of the gene-centred model of evolution, the chances of the blue-beard alleles look completely different. While a normal male will bring up an average of three sons from each litter and send into the next generation three copies of his normal variant of the gene on the Y chromosome, the Blue Beard will bring up 4 sons from each litter (which, in addition, will be well-fed) and, through them, send 4 copies of his blue-beard gene into the next population. Thus, to the contrary, it follows from the gene-centered model of evolution that the blue-beard version of the gene should prosper and should spread rapidly in the population. Thus the classical Darwinist model and the new gene-centered model of evolution yield quite opposite predictions in relation to further development of the population. They are certainly not simply a different view of a single fact and it must be accepted that only one of them can be correct.

I will not keep the reader in suspense – obviously, the new gene-centered model is correct. Even when we take into consideration that the spreading of the blue-beard alleles will lead to a gradual increase in the number of males at the expense of females in the population, so that the biological value of males will become increasingly less compared to the biological value of the rarer females, the Blue-Beard male will always pass on to the next generation a greater number of copies of his Blue-Beard alleles than a normal male will pass on copies of his normal allele. Thus the number of Blue Beards in the population will constantly increase from one generation to the next. This process will, understandably, be very disadvantageous for the population and could lead to the extinction of the particular local population or even to the extinction of the entire species. Mathematic models and data obtained in nature indicate that local populations actually do become extinct as a consequence of spreading of Blue-Beard genes.

Until now, no actual Blue-Beard gene, which would employ induction of behaviour described in the Blue-Beard model, has been observed in nature.[6] However, a number of genes achieve the same result in different ways. Blue-beard alleles in principle include t-alleles in mice and alleles of the SR system in fruit flies.[7]

For example, the t-allele of the heterozygote bearing one t-allele and one normal allele is capable of somehow damaging the chromosome with the normal allele in the future sex cells of males so that all the sperm subsequently formed with normal alleles (i.e. half of all sperm) are not viable. Thus a heterozygote male mouse has lower biological fitness than a homozygote male that has normal alleles at the given site in both of its chromosomes. It forms half as many viable sperm which, in mice, where one female is frequently fertilized in a short period of time by several males and their sperm subsequently battle for fertilization of the egg, can constitute a serious disadvantage. Simultaneously, however, all the viable sperm of a heterozygote male carry t-alleles. Homozygotes bearing t-alleles on both their chromosomes are not viable and thus it cannot happen that the t-allele would completely force out normal alleles and finally predominate in the population. Nonetheless, mathematical models have shown that t-alleles should be much more common in natural populations than is actually the case. The most probable explanation of the unexpected rareness of t-alleles in the mouse population is that a population, in which the proportion of t-alleles increases disproportionately and thus too many unviable homozygotes with two t-alleles are formed, readily becomes extinct and is replaced by a new population without carriers of t-alleles. It is apparent that these new populations can be infected at any time by an "immigrant" bearing a t-allele and would again be destined for extinction after some time. However, in the meantime new populations will be formed, whose basis will tend to consist of immigrants from the more numerous and prospering populations, i.e. from populations with a small proportion of t-alleles.

The selfish gene and the end of bared teeth and bloody talons

And now it will probably not hurt if we take a small excursion into the area of ethics. One of the reasons why a number of people in the past could not, and in the present cannot, come to terms with the validity of Darwin's theory of evolution based on selection within the species was its conclusion that evolution should basically be driven forward by a constant battle amongst the members of a single species. In this battle, the weaker would gradually fall aside and only the stronger and healthier would pass these traits on to the next generation. If man is only another member of the animal kingdom, then the same laws should also hold for him. His evolution should also be driven forward by a constant battle amongst the members of a single species, a battle which the weaker lose and the stronger and unscrupulous would win. From here, it is only a small step to the erroneous concept of **social Darwinism** celebrating strength, unscrupulousness and aggression. (Charles Darwin, of course, had nothing in common with social Darwinism, this was a sort of "creative development" of his theory.) According to social Darwinists, protection of the weak is not natural and would necessary lead to degeneration and extinction of mankind.

I don't know if it is necessary in the present day to explain why these concepts are erroneous. Perhaps, just briefly – whether or not any type of behaviour is ethically permissible is not decided by whether it is natural, i.e. whether it is practiced or not practiced by our animal relatives or whether it was practiced by our evolutionary ancestors. Thus an evolutionary biologist cannot give a qualified opinion or even somehow make a decision on the moral permissibility of a certain kind of behaviour but, at a theoretical level, this is always and only the domain of the **philosophical discipline of ethics** (in practice, of course, far more likely the opinions of recognized moral authorities at the given time, or even more likely, ingrained habits). It is not important whether these ethics have their foundations in religion or, say, Kant's moral imperative to behave in the way that we would want the majority of mankind to behave. Darwin's theory can maximally warn us against ourselves – careful, being good is probably not part of our basic nature and thus it definitely need not be right to subject our behaviour to our natural inclinations.

Although the examples on the basis of which I have so far explained the foundation of the theory of the selfish gene (i.e. the theory of interallelic selection) may not, at first glance, be very convincing in this respect, in actual fact the new theory substantially weakens the theoretical basis for social Darwinism. Indeed, the theory of the selfish gene indicates that the particular interests of an individual, specifically an "attempt" to maximize his biological fitness, are not the main criteria in evolutionary success and the main subject of interest of biological evolution. An attempt to maximize biological fitness is only a side effect of attempts of the individual alleles to maximize the number of their own copies passed on to the next generation. The individual alleles achieve this in different ways, in some cases even at the price of reducing the biological fitness of their carriers. This reduction need not take the form of killing one's own daughters or programming genes on sister chromosomes to damage future sperm. In a great many cases, to the contrary, alleles program their carriers to behave altruistically towards their surroundings, i.e. to help other members of the population, even at the price of reducing their own biological fitness. The observation of William Hamilton that, when considering the selection advantageousness or disadvantageousness of a certain trait, for example a certain pattern of behaviour, it is necessary to study, not the "exclusive" biological fitness of the bearers of the particular behaviour, but its "inclusive" fitness, formed the basis for the theory of interallelic selection (the theory of the selfish gene). The **exclusive biological fitness** expresses how the given trait affects the number of offspring that its **bearer** produces during his lifetime, more exactly, how it affects the number of these offspring that live to adulthood and successfully reproduce themselves. In contrast, the **inclusive biological fitness** expresses how the particular trait affects the number of offspring that the **bearer and his blood relatives** produce during their lifetimes.[8]

From the standpoint of biological fitness, one's own progeny are, of course, more valuable, as each of them bears half of his genes (more exactly, copies of his alleles). The progeny of his siblings each bear only one quarter of his genes and thus, from an evolutionary standpoint, have only half the value of his own children. Even so, from an evolutionary standpoint it is more advantageous for an individual if he saves the lives of three nephews or nieces than one brother (or one son), or if he sacrifices his own life to save the lives of three brothers. If his sacrifice were to save the lives of only two of his siblings (and he was still capable of reproducing at that time), he could suffer from an evolutionary point of view, because the number of persons that, in actual fact, have other fathers than is given in their birth certificates is disturbingly high even in our apparently monogamous society. Hamilton demonstrated that parental and sibling altruism and a number of other patterns of altruistic behaviour known in nature, whose formation was very difficult to explain on the basis of classical Darwinist evolution based on selection within the species (individual selection), can be explained relatively easily on the basis of selection occurring at the level of groups of genealogically related individuals i.e. **kin selection**. Here, the main criterion for the advantageousness of the spreading of certain allele is its effect on the inclusive biological fitness of its carriers. Dawkins subsequently demonstrated that the best way of estimating the evolutionary advantage or disadvantage of behaviour towards individual members of the population in individual cases is to imagine that we are an allele on a chromosome of an individual and that, on the basis of statistical considerations and the information that we have available, we attempt to decide the probability with which our own copy will occur in the genome of a certain member of the population. If we are a gene on an autosome, then there is about 50% probability that our own copy is located on an autosome of our siblings (it can be more because the relevant allele could be very common in the population – however, an allele in the genome has no way of finding this out. But it can also be less – here I would like to loosely cite my colleague Zrzavý: it depends on whether that man that, together with my brother, we call daddy is actually the biological father of the two of us.)

Box 8.7 Autosome

Autosomes are all the chromosomes with the exception of sex chromosomes. While Y chromosomes occur in each generation only in the bodies of males and X chromosomes spend twice as much time in the bodies of females than in the bodies of males (because there are two copies in each cell), autosomes occur with the same frequency in males and females.

Box 8.8 Do psychological factors or genes determine human behaviour?

Basically, actually genes – although rarely in that they would directly affect our conscious and subconscious processes of evaluating information and thus determine our behaviour. However, they controlled the creation of our bodies, including our brains, and thus predetermined how our brains will respond to various stimuli that come to them through our senses. The traditional differentiation of nature vs. nurture is thus, basically, artificial – the behaviour of people is mostly determined by what they learn, i.e. culture; however, what we learn is predetermined by our genes.

Genes can discover whether they are present in the organism of a male or female and, according to this, affect the behaviour of the individual. However, if they are in the body of a different individual (e.g. brother or sister), they can only "guess". Imagine that we are a gene on chromosome X and that we are located in the organism of a woman. If we are a gene from the X chromosome of our mother, there is 50% probability that our copy will exist in the genome of any brother or sister. However, if we are on an X chromosome that was derived from our father, then there is 100% probability that our copy is present in the genome of any sister (the father has only one X chromosome and thus could only pass its copy on to his daughters) and also 0% probability that our copy could exist in the genome of any brother (men inherit their X chromosomes from their mothers and Y chromosomes from their fathers). Thus an average allele on an X chromosome of a woman can "expect" with 75% probability that its copy is present in the genome of any sister and only 25% probability that it is present in the genome of any brother. On this basis, it could be expected, for example, that women will help their sisters more than their brothers. If we are an allele of a gene on a Y chromosome, then there is 100% probability that our copy is present in the genome of our brother. In contrast, there is 0% probability that our copy will be present in the genome of a sister. Thus, theoretically, brothers should behave with greater altruism towards one another than towards their sisters. However, there are far fewer genes on the Y chromosome of human beings than on the X chromosome. Thus, it is not entirely certain that the interests of genes on Y chromosome of a male can predominate over the interests of his genes on an X chromosome (which can "expect" with 50% certainty that they occur in the genome of any brother or sister, but which spend twice as much time in the genomes of females, see Box 8.7). Thus it is quite possible that brothers will help sisters more than brothers. I will leave experimental verification of the theoretical conclusions to sociobiologists who will be willing and capable of immersing themselves in records of probate proceedings or will ask the customers in toy shops for whom and at what price they are purchasing gifts. I certainly hope it works out the way I derived it. If not, it can't be helped – the behaviour of human beings is affected by a great many factors, including psychological factors and, in a particular case, some of them can be stronger than manipulation on the part of genes for altruism located on X and Y chromosomes.

Summary and incitement

The main message of this chapter should thus be that classical Darwinism and Neodarwinism, based on competition of individuals within a species for the greatest biological fitness, have become outdated, as most evolutionary biologists base their considerations on the theory of the selfish gene. This theory states that, in the case of sexually reproducing organisms, the driving force for evolution is competition between various variants of a single gene for passing on the greatest number of its own copies to future generations. Only in some cases is this competition between genes accompanied by competition for the greatest biological fitness of individuals within a species. When compared with the Darwinist theory, the Dawkins-Hamiltonian theory of the selfish gene explains a greater proportion of biological phenomena, including the formation of some types of altruistic behaviour.

But beware! In the next chapter, we will show that Dawkins correctly felt out the basic inadequacy of the Darwinist theory based on the competition within a population for the greatest biological fitness; however, the solution that his theory of the selfish gene offered is also most probably erroneous. We will show that even his evolution of adaptive traits based on competition of alleles within a single locus cannot function in sexually reproducing organisms.

Footnotes

1. The main idea that later led to the selfish gene theory, was published by W.D. Hamilton in the *Journal of Theoretical Biology* 7: 1–16 and 17–52 in 1964. As far as I know, this is actually an excerpt from his doctoral thesis, which the author had considerable difficulty defending, as so often happens with fundamental discoveries.
2. While I stated in the footnotes above that the reader should also read a number of works, I think that the reader definitely must read *The Selfish Gene* by Richard Dawkins (Oxford University Press, Oxford, 1976). In fact, all the books by this author are worth reading. (And I say this quite seriously, and not because I want to smooth things out with him because of the indisputable fact that I am going to bring into discredit his theory of the selfish gene in the subsequent chapter.)
3. A number of important articles have been published on the subject of underestimation of the scientific erudition of the authors of popular books. Unfortunately, in my database, I found only an article devoted to S.J. Gould, *Social Studies of Science* 32: 489–524, 2002.
4. The parasitic liver fluke doesn't seem to be a good example at first glance. A large percentage of parasitic organisms are capable of asexual reproduction. However, in all cases, this is secondarily formed asexual reproduction – the ancestors of the particular parasitic species reproduced sexually. The main reason for transition to asexual reproduction is apparently an "attempt" to produce, within the host, genetically identical progeny and thus prevent selection within the population located in a single host (called the infrapopulation). This selection would mostly lead to an increase in the virulence in the infrapopulation, the host would be too seriously damaged by the parasites and this would lead to a reduction in the basic reproduction constants for the particular parasite. The basic reproduction constant is the most important property (criterion of biological fitness) of a parasitic species or parasitic population. It expresses the number of new hosts that are infected, on an average, by one infected host (in a population that has not yet come into contact with the given parasite, i.e. a population where there are no previously infected or immune individuals). The evolutionarily most successful parasite species have been those that reproduce within the host asexually (and thus cannot develop towards a lower reproduction constant) and sexually produce only the stage that left their hosts for the external environment and act as a source of infection for other individuals of the host species.
5. See: *Science* 156: 477–488, 1967.
6. A number of cases are known in nature where the sex ratio of progeny is adjusted through cannibalism. How-

ever, I think it has never been demonstrated that this constitutes a manifestation of the Blue Beard gene. For example, in some species of parasitic hymenopterous insects (*Copidosoma floridanum*), in which mating occurs between siblings and where one brother can theoretically manage to fertilize all the sisters, a specialized type of larva kills the larvae from which males would develop. The killer larvae also finally die and their main (and perhaps only) task is to ensure a shift of the sex ratio in favour of females *Nature*: 360: 254–256, 1992.

7. You can read about the SR alleles of fruit flies, e.g., in *Heredity* 83: 221–228, 1999, and about the t-alleles of mice, e.g. in *Trends in Genetics* 14: 189–193, 1998, *Evolution* 50: 2488–2498, 1996.
8. The relevant relationship is very simple: altruistic behaviour of individuals is worthwhile if $rb > c$ (r – relationship of mutually assisting individuals, basically indicating how much larger will be the probability that an allele of the gene for altruistic behaviour will be shared by two specific individuals than that it will be shared by two randomly selected individuals in the population, b – the advantage that altruistic behaviour provides to the assisted individual and c – the cost that the assisting individual must pay for assisting. The price and the advantage are measured in terms of the relevant change in the biological fitness.). *Journal of Theoretical Biology* 7: 17–52, 1964.

Chapter 9 It was only a matter of time or the skeleton in the cupboard of the selfish gene theory

The selfish gene theory has become a standard starting point for discussions of evolution and, in this role, has almost completely displaced Darwin's original theory of individual selection. When a modern evolutionary biologist looks for the biological importance (purpose) of certain properties of organisms, for example, a particular organ or a certain pattern of behaviour, he does not usually ask how this property increases the biological fitness of its bearer. The first thing he considers is how the particular property increases the chance that a copy of the gene that determines this property of its bearer will be transferred to the next generation.

As the selfish gene theory was established only 30 years ago, it has not yet found its way from the workplaces of scientists studying evolutionary biology to the pages of basic textbooks in the field and even less into general awareness. This is fully in agreement with the "Structure of Scientific Revolutions" of Thomas Samuel Kuhn (see Box 3.7 on p. 43). Until the old proponents and co-creators of the original Neodarwinist theory leave for a better world or at least for a well-deserved rest, and are replaced by a new generation of scientists that, metaphorically speaking, absorbed the selfish gene theory with their mother's milk, and until these scientists decline sufficiently to start writing their own textbooks, the new theory will be presented in textbooks maximally as a sort of cream on the cake and not as the basic theoretical framework of the field. And, as I shall try to convince you in this chapter, it is not certain that such a situation will ever occur for the selfish gene theory.

Everything is wrong, let's go back to the trees

It is quite possible that the selfish gene theory is not as impregnable as it seemed to be to its creators and as it appears at first glance. I am of the opinion that the selfish gene theory (i.e. the theory of interallelic selection) only seems to solve the problem for which it was originally created and that it is, similar to Darwin's older theory of individual selection, fundamentally erroneous.

It should be borne in mind that the selfish gene theory was created to explain the functioning of natural selection, and thus also the evolution of adaptive traits (useful characteristics), amongst species that reproduce sexually, i.e. under conditions where there is limited heredity of biological fitness. The biological fitness of an individual is determined by the set of his characteristics, i.e. his phenotype. On the one hand, this phenotype is determined by the environment in which the individual developed and in which he lives (e.g. the amount of food that was available for his development) and, on the other hand, by his genotype – the combination of his alleles. However, the genotype is not created by simple copying of the genotype of one of his parents, but by random mixing of half of the alleles of one and half of the alleles of the other parent. The heredity of the biological characteristics of an individual thus slowly disappears, the good and bad qualities are transferred to the descendants and to their descendants to an ever decreasing degree (as the alleles that are responsible for them are diluted). The solution offered by the selfish gene theory consisted in focusing our attention on the behaviour of the individual genes; to be more exact, on the individual alleles in the framework of the genome. Alleles are formed by only a short continuous segment of the DNA on a chromosome. Thus, they cannot be affected by genetic segregation and can only with low probability be affected by genetic recombination (a description of both these processes is given in Chapter 3). Consequently, they are almost always transferred from one generation to the next in unaltered form. As they are transferred from one generation to the next by copying and are not created anew in each generation like the genotype is, they can become the subject of natural selection and thus the subject of evolution of adaptive traits.

However, there is one difficulty which the author of "The Selfish Gene" only seemed to resolve. Alleles do, in fact, multiply by copying and are transferred between generations in an unaltered form; however, in each generation they become part of a different genotype – each time they find themselves in a different association of a completely different set of the alleles of the other genes. **Moreover, the effect of the individual alleles on the phenotype of the host is almost always dependent on the other alleles of the other genes that are also present in the genotype of the individual.**

The vast majority of genes affect the quantitative traits. A single allele can affect a particular trait positively in one case (e.g. can cause elongation of the limbs during the growth of an individual) and, in another case, **in the context of a different genotype** (i.e. in combination with other alleles), need not be manifested (its bearer may have limbs of average size), or could even have a negative effect on the same trait – its bearer could have shorter limbs. Physicians are well aware that the clinical manifestations of a single gene causing a particular genetic disease differ dramatically even amongst members of a single family, i.e. even in a situation where the individual patients bear the same copy of a certain damaged gene and, because of their mutual relationship, their genotypes are rather similar.[1] Amongst other things, this draws attention to another problem.

Box 9.1 Quantitative and qualitative traits

Only very few of the traits that are exhibited by organisms are qualitative, i.e. have the nature of being "all or nothing". The presence or absence of a certain mark on the surface of the body could be an example of such a trait. The size, intensity or colour of the mark, similarly to the size of the body and its parts, however, are quantitative traits, i.e. we can measure the intensity, size or, in relation to behaviour, the probability or frequency of its occurrence. While qualitative traits can be determined by the presence or absence of certain (frequently dysfunctional) alleles of one gene, quantitative traits are usually dependent on, or are at least affected by, a large number of genes located in various parts of the genome.

Not only is it true that a single allele can have a positive effect on a certain morphological or physiological trait or on certain behaviour in one case and a negative effect in another case, depending on the overall genetic background; in addition, it also holds true that a **single trait can sometimes have a positive effect and sometimes have a negative effect on the biological fitness of its bearer**. It not only holds that the effects of certain alleles on the phenotype of the individual are generally dependent on the presence of other alleles (genetic background) but, in addition, the effect of the individual traits on the biological fitness of the individual is usually dependent on the presence of other traits.[2]

However, this leads to the following considerations: in spite of the fact that the individual alleles are transferred between generations in unaltered form, the biological characteristics of organisms are inherited only to a very limited degree and overall biological ability is inherited to an even lesser degree. We, of course, know of traits whose manifestations are controlled by a single gene and that have the same or almost the same effect on biological ability for any host. However, even here, it holds in general that the biological fitness of an individual depends on whether the individual bears one or two copies of the given allele in his genome. In addition, there is only a small percentage of these traits and the fact that textbooks describe at least some of them is most probably a result of the fact that, by chance, relevant alleles of other genes are present that would be capable of modifying the effect of the studied alleles on the given trait in the studied population, or that the research worker carefully eliminated the bearers of these alleles from the test group in advance.[3] (I tactfully exclude the other possibility, i.e. that the research worker would not notice the effect of other genes or that he would perhaps only pretend not to notice them.) It is well known that even the great Johann Gregor Mendel spent a number of years in selecting suitable, i.e. sufficiently stable and sufficiently "orderly" behaving traits, before he carried out his famous experiments through which he demonstrated the "hard" heredity of biological properties. It was this choice of suitable traits (as is now apparent, usually dependent on complete destruction of part of a gene and thus creation of a completely dysfunctional protein product of this gene) that played a key role in the success of his study. If he had made a random selection of traits for his experiments, he could well have reached the completely opposite conclusion – he would have discovered that heredity is "soft" and gradually vanishes.[4]

How not to train a rowing club

How did Richard Dawkins come to terms with the dependence of the manifestation of the individual alleles on the genetic backgrounds and with the dependence of the effect of the individual traits on the biological fitness in the presence of other traits? Elegantly, but not, in my opinion, entirely cleanly. For this purpose, he employed an example from the world of sports, which I shall attempt to freely retell. Imagine that we have the coach of a rowing club, who is to prepare his team for a major eights competition. He can choose from a large number of candidates, and more are constantly moving up from the junior team. The trainer originally worked as an evolutionary biologist, so he employs a simple method reminiscent of natural selection. He places all the candidates in boats and holds a race. He writes down the order of the boats at the finish and the compositions of the individual crews. He sends the competitors in the last boat home and calls up candidates from the junior team in their place. Then he breaks up the individual crews and creates completely new crews by random combination of the members of the club. Then he starts another training race. Dawkins tells readers that, in this untraditional way, the trainer would gradually select a team of rowers that would not only be strong and fast as individuals, but would best be capable of applying their capabilities in the environment of the particular team – in cooperation with the other members of the team. Hmm. I have the feeling that colleague Dawkins was very lucky that rowing trainers do not read books on evolutionary biology. In the opposite case, I can quite clearly imagine that a group of angry trainers would chase him through Oxford, waving broken oars, throwing hard objects and shouting words that an Oxford don would not normally encounter on university soil. And quite rightly so, in my opinion. This training method would hopefully remove the clumsiest men from the club, persons with serious physical and mental handicaps (such as individuals who are good rowers, but sometimes fall asleep during a race). However, in no case could it ensure an improvement in the performance of the club or acceptance of the best members of the junior club. The reason is, I think, quite obvious: the abilities of the individual rowers are not additive from the standpoint of the quality of the entire team, but affect one another positively and negatively. If, for example, an unusually strong rower moves up from the junior club and is randomly placed in a team of average quality, his presence need not improve the performance of the team and need not improve the position of the boat in the race. It is quite possible that the performance of the team will decrease – if he doesn't directly knock the oars out of his colleagues hands, he will at least cause the boat not to move in a straight line in the race. (Now, I can quite imagine how a group of angry trainers would chase me through Albertov in Prague, waving broken oars, throwing hard objects (I am not sure about Oxford, but there is certainly no lack of such objects in Albertov, which is constantly in disrepair) and shouting words that a professor at Charles University would not normally encounter in a university environment.) Okay, in actual fact the presence of an excellent competitor amongst competitors of average quality would have a somewhat different effect, but the results would be similar. The average performance of the club would probably not decrease over time, but would probably also not improve and, in any case, the best members of the junior team would not be accepted (with their extraordinary performance, they would not fit into a team with normal performance). For completeness, I must admit that, as far as this last statement goes, Dawkins also did not suggest otherwise. (Now I can imagine Richard Dawkins chasing me down Albertov in Prague, waving a broken-off display from a notebook, throwing hard objects and shouting words that a professor at Charles University would not normally encounter, at least not in Oxford English.) The main reason why the performance of the team could not improve is that the effect of the characteristics and abilities of an individual rower on the performance of the team is affected or even directly determined by the characteristics and abilities of the other members of the team and the **frequency, with which the individual types of competitors occur in the rowing club**.

Neither Darwin nor Dawkins, so what now?

Did the previous sentence about frequency remind you of anything? It should. Remember the chapter in which we discussed frequency-dependent selection. If the effect of a certain allele on the biological fitness of an individual is dependent on its frequency or on the frequency of other variants of genes in the population, then the effect of the evolutionary battle cannot be decided on the basis of the result of simple comparison of the average biological fitness of its bearers, but only on the basis of the results of analysis based on game theory. And this analysis indicates that **the result of the evolutionary battle is decided, in the last instance, not by which allele ensures greater biological fitness for its bearers, but which is connected with an evolutionarily stable strategy**, i.e. a strategy that, when it once predominates in the population, cannot be forced out by another strategy. Thus, similar to the model of the "dove and the hawk" – in which neither the dove nor the hawk finally wins out and an equilibrium is established between the two strategies – in a great many other cases, an allele ensuring its bearers the greatest biological fitness often cannot win. Consequently, the population constantly remains polymorphic in a great percentage of its genes (a figure of 15–50% of genes has been given[5]), i.e. the individual genes persist in the population in a great many variants. Each of these variants is advantageous for its bearer in combination with the other alleles of that gene and with other combinations of alleles in the other genes. On the one hand, this ensures that a single allele cannot completely predominate in a certain site in the chromosome (at a single locus) and simultaneously this means that the alleles of the other genes will be associated with other alleles in each generation, forming a different, in each case completely unique genotype of the individual. Let's go back to the beginning. As a consequence of the fact that the allele is present in the company of different alleles in each generation, in a differently mixed genotype, its effect on the phenotype of the individual will be different in each generation and its contribution to the biological fitness of the individual will also be different. However, this means that neither **classical individual selection, which forms the basis for classical Neodarwinist theory of evolution, nor interallelic selection, on which the theory of the selfish gene is based, cannot work effectively in sexually reproducing organisms.** Thus, in the last analysis, we find ourselves in a similar situation to that in which Darwinists were before the discovery of hard heredity of biological traits. The heredity of traits dependent on the effect of many genes and their mutual interactions, i.e. the heredity of the vast majority of traits, is soft for sexually reproducing organisms and slowly vanishes from generation to generation (as the combination of alleles determining the individual forms of traits become more distant from one generation to the next). Consequently, the only known mechanisms of evolution of adaptive traits, Darwinian individual selection or even Dawkinian interallelic selection, cannot function in these organisms. However, organisms clearly exhibit adaptive traits and this is frequently very obvious. Where do these traits come from? Or where did we make a mistake in our considerations? Or did the nasty author intentionally mislead you with false proofs? (And did he actually manage this, the scoundrel?)

Summary and incitement

To summarize. In this chapter, I attempted to demonstrate that the solution to the problem of the impossibility of Darwinist evolution in sexually reproducing organisms proposed by Dawkins cannot function in actual fact. Dawkins correctly recognized that, in contrast to the genotype that is not inherited in sexually reproducing organisms, but is created anew in each generation by mixing the alleles of the parents, the individual alleles are inherited from one generation to the next mostly in unaltered form. However, I am of the opinion that he did not sufficiently realize that, because of genetic polymorphism, gene interactions and the dependence of the biological fitness corresponding to particular alleles on their frequency in the population, the effect of the individual alleles on the phenotype of their bearer and on his biological fitness changes from one generation to the next. Because of these changes, a single allele is advantageous for its bearer in one case and, to the contrary, disadvantageous in another case, so that it cannot be fixed in the population or, on the other hand, be completely removed from the population. Thus, in sexually reproducing organisms, neither Darwinian individual selection nor Dawkins – Hamiltonian competition of alleles can function. The genetic data that form the basis for the conclusions in this chapter are discussed in greater detail in the next chapter. I would like to warn my readers in advance that this chapter will be somewhat more difficult to understand for the non-biologist. However, in the worst case, it can be omitted entirely or at least its first half can be omitted without great detriment. But why? No one is going to examine you on its contents and no-one is likely to return you your money for unread chapters.

Footnotes

1. The aspect of various manifestations of the same gene for hereditary diseases in various persons is discussed, e.g., in the review in *Hereditas*, 125: 1–9, 1996.
2. For example, the mutation causing α–thalassemia is harmful for its host, as a large part of the haemoglobulin in the red blood cells will be present in the form of the not-very-functional homotetramers γ4 and β4. However, if an allele with this mutation occurs in persons with β–thalassemias, it prevents the formation of the poorly soluble homotatramer α4 in this person, substantially reducing the clinical manifestations of β–thalassemia and thus increasing the chances of survival for its host. *British Journal of Haematology* 52: 465–473, 1982, *British Journal of Haematology* 53: 411–416, 1983.
3. For example, the wrinkled shape of pea seeds, i.e. one of the traits whose heredity was studied by Mendel, is caused by the insertion of a foreign element (a transposome) into the gene for the enzyme required for synthesis of the starch molecule. As a consequence of this insertion of the foreign DNA into the middle of the gene sequence, the allele with transposome codes a non-functional protein *Cell* 60: 115–122, 1990. The situation is also similar for other classical studies traits *Proceedings of the National Academy of Science, U.S.A.* 94: 8907–8911, *Science* 277: 1611, 1997.
4. For persons with a more serious interest, I can recommend (if not for reading, then at least for perusal) the book by Ernst Mayr *The Growth of Biological Thought*. Cambridge: The Belknap Press of Harvard University Press, 1982.
5. In actual fact, there will be far more polymorphic genes. The original estimates are based on data obtained using allozyme analyses, which permits differentiation only of proteins differing in their mobility in an electric field and, in addition, only those genes whose most common allele occurs in less than 90% of individuals are classified as polymorphic.

Chapter 10 Theoretical discussions are all very well, but what does the green tree of life have to say about all this?

I could probably not make a living as a politician, con artist or attorney; nonetheless, I have the feeling that, if we want to deceive someone (or even the general public) in a discussion, it is probably better to base this on false basic data rather than to try using false arguments to draw erroneous conclusions from correct data. Thus, it is worth considering the validity of the data that formed the basis for my rejection of the validity of the theory of the selfish gene in the previous chapter. I must admit that I am not a hundred percent certain (and perhaps not even 90 %). However, I tend towards the opinion that these data are correct and, in the following chapters, I will attempt to demonstrate some reasons that led me to this conclusion. Still, as I said, I cannot be completely certain. It will probably be better if those with "black-and-white" thinking leave out the rest of this chapter, or at least go straight to the section called "And why not simply test it?" The manufacturer gives this warning: the following passages contain a substance that could cause the consumer a lack of appetite, headaches and serious health problems. Addictiveness and carcinogenicity have not yet been demonstrated; however, the manufacturer refuses any responsibility in this respect.

How many genes fit onto the head of a pin and how many are required for one trait?

Where could the hitch be? Primarily, in my statement that a great many genes participate in the creation of most traits, where the effects of the individual alleles are not simply added together in the final result (i.e. their effects are not additive). If this were not true, a substantial part of my argument would be invalid. (My objections to the theory of the selfish gene would, however, not be completely invalid, although I would have to base them on the statement that the effects of the individual traits on the biological fitness of an individual cannot be simply added together.) What do we really know about the usual number of genes that participate in the formation of a single trait? How frequently is the final form of a trait (eye colour, resistance to tuberculosis) determined by a single gene and how frequently are several genes, or even a large number of genes involved? In the case of traits determined by several genes, how frequently are the effects of the individual genes simply added together and how frequently are there more complicated relationships amongst them, in which a single allele in the presence of a single allele from another gene shifts the character of the trait in one direction (i.e. increases resistance to tuberculosis) and, in the presence of another allele, in quite the opposite direction?

The results of genetic experiments in this area are, to say the very least, certainly not unambiguous. In some cases, we really encounter traits whose form is controlled by a single gene. These tend to be traits of a qualitative nature, for example, a trait consisting of the presence or absence of some structure, or of some protein in the cell. In such a case, more detailed analysis usually shows that a damaged, and thus dysfunctional, allele of the relevant gene is responsible for the particular trait. However, it is necessary to be aware that such a dysfunctional allele does not constitute the evolutionarily original variant of the particular gene, but rather a variant that was formed secondarily, as a consequence of damage to some functional alleles. It is thus, at the very least, doubtful whether study of the behaviour of these dysfunctional alleles can tell us anything substantial about the character of biological evolution.[1]

The vast majority of biological traits are of a quantitative nature and are affected by a larger, frequently very considerable number of genes.[2] However, genetic analyses simultaneously demonstrate that, as a rule, a few genes affect the particular trait very strongly and a great many more genes affect it only very little – together they explain only a small fraction of the genetically determined variability. However, the problem with these studies is that they are performed on relatively small model populations with low genetic variability. It is quite probable that the same study performed on a different model population would again demonstrate the presence of several principal genes and a great many auxiliary genes (that affect the particular trait very little). However, in a number of cases, the gene that manifests a strong effect in the first study would fall well within the category of weak genes in the second study and an originally weak gene from the first study would appear as a strong gene in the second study.[3] This fact disturbs the sleep of a great many immunologists searching for the basis for resistance to parasites. The genetic systems described in one strain of laboratory mice could look completely different when studied in a different strain. Scientists generally resolve this in the simplest possible way: they perform their studies exclusively on the same strains as their predecessors used. In this way, they make certain that the results will be compatible with the previously published data. Simultaneously, it is known that a great many strains of laboratory mice are quite related and are derived from only a few mouse lines. Thus, it can be expected that the genetic variability of natural populations of mice and thus also the variability of their genetic systems determining the individual forms of traits will, in fact, be substantially greater.

Experimental evidence for the multigenetic determination of traits and simultaneously for the great plasticity of genetic systems has been available since the 1920s. One of the greatest personages of evolutionary biology, the important statistician and geneticist, R.A. Fisher, demonstrated that the manifestations of even very strong genes can be substantially affected by the presence of other genes. For example, if a new mutation appears in the population, it usually has the greatest phenotype manifestation in the first generation and, over the next few generations, the phenotype of the carriers of the particular mutation gradually returns to normal. Finally, the carriers of the mutation cannot be differentiated from the other members of the population on the basis of their phenotype. During the process, nothing happens to the actual mutated allele and it is present in the population in the same form as when it originally appeared. This can be verified by crossing the carriers of the mutation with the members of another population in which the mutation did not previously occur. Surprisingly, the mutation begins to appear with its original strength in the new population. Fisher correctly explained the gradual reduction in the phenotype manifestation of a mutation as the result of the accumulation of **modifier genes** (more exactly modifier alleles) in the gene pool of the monitored population. Most mutations with a major phenotype effect reduce the biological fitness of their carriers. This is understandable – the phenotype of an organism has a long period of gradual improvement behind it and a random change will thus tend to be a change for the worse. As a consequence, in a population containing a large number of a certain mutation with strong expression, those alleles that act to modify (or even completely neutralize) the effect of the particular mutation on the phenotype of its carriers will have a selective advantage. These alleles from various genes will gradually accumulate in the population and finally their presence in all the members of the population will lead to complete disappearance of the effects of the mutated alleles.[4]

Two and two equals minus seven or about genetic interactions

The question of whether the effects of the individual mutations on a phenotypic trait are usually simply added together, or whether the relationships between the individual genes are more

Box 10.1 Genetic interactions

Genetic interaction is the dependence of the degree or character of the manifestation of one allele on the presence of other alleles. If these are alleles of a single gene, only several basic possibilities can occur in diploid organisms that bear two alleles from each gene. If allele A completely suppresses the manifestation of allele B and an individual with two A alleles then looks the same as an individual with one A allele and one B allele, then allele A is denoted as **dominant** with respect to allele B, while allele B is denoted as **recessive** with respect to allele A. For example, this is the case of the allele for brown and blue eye colour – a homozygote with two alleles for brown eye colour does not differ from a heterozygote that bears one allele for brown colour and one allele for blue colour (simplified somewhat – I hope experts will forgive me). If the expression of alleles A and B are averaged out and an individual with a pair of alleles AB (heterozygote) has traits somewhere between the traits of individuals with two A alleles and the traits of individuals with two B alleles, this is called **semi-dominance** (incomplete dominance) – the allele of the pair that is manifested more strongly in the heterozygote is denoted as semi-dominant. For example, semi-dominance is exhibited by the S allele responsible for the detrimental manifestations of sickle cell anemia – a homozygote with two alleles is much worse off than a heterozygote with one normal and one S allele. The case when a heterozygote with a pair of alleles AB exhibits the relevant trait to a greater degree than either homozygotes AA or BB is termed **super-dominance**. Only a limited number of types of gene interaction can occur between the alleles of a single gene in diploid organisms. There are a much greater number of types of gene interactions between the alleles of various genes, called **epistatic interactions**, because the number of interacting alleles can attain any number. For example, a particular allele of one gene can directly affect the manifestations of alleles A and B of a different gene and can also affect whether allele A will be dominant or recessive towards allele B.

complicated, i.e. to what degree **genetic interactions** play an important role in the formation of traits, remains unresolved. In principle, genetic interactions can also be considered to include interactions between two alleles of a single gene in diploid organisms. This type of genetic interaction, specifically relationships of the **recessivity, dominance and super-dominance** types, is quite common in nature.

If an individual carrying a new mutation is crossed with a normal member of the population, the phenotypic expression of the mutation in the offspring carrying normal alleles and one mutated allele are usually less than they would be in an offspring carrying two mutated alleles, i.e. the mutated allele behaves as a semi-dominant allele. In contrast, if two carriers of two different alleles that normally occur in natural populations are crossed, the phenomenon of dominance and recessivity is encountered more frequently. The expression of one allele completely suppresses the expression of the second allele. Therefore, homozygote individuals carrying two dominant alleles in their genomes cannot be differentiated according to their phenotype from heterozygote individuals that have one dominant and one recessive allele in their genomes. A situation in which the phenotype of a heterozygote does not lie somewhere between the phenotype of a homozygote carrying two identical copies of one or the other allele, but is quite different, is also encountered quite frequently. For quantitative traits, such as body weight, we frequently encounter super-dominance in this case, where heterozygotes are larger (or smaller) than any of the homozygotes. The difference between the behaviour of new mutations (quite frequently additivity in the effects expressed as a result of incomplete dominance) and the behaviour of normally present alleles in the population is apparently caused by the **accumulation of modifier genes** that act to suppress the expression of one of the two alleles in the presence of another allele. Amongst other things, this is a further demonstration of the plasticity of genetic systems and the complexity of the genetic architecture of phenotype traits. It also shows that, as a general rule, a great many genes cooperate in the creation of the final form of the individual traits, where some of them do not affect any trait directly, but rather indirectly in that they modify or completely change the relationships between other genes (or alleles) (e.g. from partial dominance to dominance or to super-dominance).

Interactions between various genes (and not between the alleles of a single gene) are called **epistatic interactions.** Although these relationships are frequently a subject of interest for theoretical geneticists and practical breeders, their actual significance in nature is not very clear at the present time. The weight of the body of an individual is a trait that is of considerable interest to breeders. This trait is usually subject to the interaction of a great many genes, but the effects of these genes are very frequently additive. Far less research has been done on the genetic architecture of other types of traits. However, it has been found that a number of more complex genetic interactions are involved here than in the case of body weight. Studies concerned with the genetic architecture of traits directly connected with biological fitness, such as the number of offspring, showed that the number of genes affecting the given trait additively is usually approximately half as many as those participating in epistatic interactions. Simultaneously, the way in which the relevant experiments are performed reduces the chance of discovery of any interactions and underestimates their strength and thus also their importance. Here, we ignore the fact that the studied population corresponds to only a small fragment of the overall genetic variability of the population, so that the less common alleles that can have a very strong effect in genetic interactions are not present here at all. In order for it to be possible to discover the presence of interactions in which a larger number of genes participate simultaneously, a substantially greater number of studied individuals would be required than the number that is quite sufficient for determining the effects of the individual genes or for demonstrating interactions between two genes. The necessary number of individuals increases very rapidly with the number of genes participating in the given interaction. Thus, in order to demonstrate a strong interaction of two genes, it is sometimes sufficient to have an experimental set consisting of 100 individuals; in order to demonstrate a similarly strong interaction of three genes, a set of more than 1 000 individuals would be required. If, before commencing a study, a scientist were to perform an analysis on the strength of the study, a power analysis, he would mostly discover that the size of his test set would be more or less sufficient for discovering the main effects of the individual genes and interactions between pairs of genes; however, it will almost certainly not be possible to either demonstrate or exclude the effects of higher order interactions.[5]

The last and apparently the most important reason why our experiments would tend most probably to underestimate the number and importance of interactions in the formation of phenotype traits again consists in the methodology employed and simultaneously also in the usual interpretation of the results obtained. The statistical methods employed to search for interactions reveal statistical interactions.

Although not all biologists are aware of it, the word interaction has a somewhat different meaning in statistics than in genetics and the strengths of statistical interactions cannot directly or simply indicate the strengths of interactions between the studied genes. Imagine that we have two genes, the first with alleles X and x, and the second with alleles Y and y. Allele X in the presence of allele Y causes an increase in body weight by 10% (compared to allele x) and allele Y

Box 10.2 Power analysis

This is a statistical method that allows us to estimate the number of objects that must be studied in order to be able to statistically demonstrate a certain dependence. Basically, it can be stated that quite negligible and, from a practical standpoint, completely uninteresting dependences can be demonstrated in a sufficiently large set. For example, in order to demonstrate the effect of the shadowing of a field by flying swallows on the crop yield, all the arable land in Europe would probably have to be reserved for our "very important" study. Before a scientist decides to perform a more demanding study, he should first estimate, on the basis of the available information and using power analysis, how large a test set he will require to have a reasonable chance of demonstrating the studied effects. Power analysis simultaneously permits estimating *post factum* how far we can believe the results of a study that did not demonstrate the existence of the studied dependence.

Box 10.3 Statistics

For the natural scientist, statistics is primarily a set of mathematical procedures that allow him to search for laws in a world in which the element of chance is constantly in effect. Most frequently, the use of statistical methods consists in testing the validity of hypotheses. If, for example, we find that 20 students infected by the *Toxoplasma* protozoa are, on an average, taller than 72 uninfected students, the relevant statistical method allows us to estimate, on the basis of the heights of all the 92 students, the probability that the observed difference in the average height of the infected and uninfected students is only a matter of chance. In this case, the t-test told us that this probability equals only 2.6%, indicating that there is great probability that there is some dependence between the height of the students (men) and infection by *Toxoplasma gondii*. However, the results of statistical tests understandably cannot answer the question of whether the infection increases the growth of students or whether taller students have a greater probability of becoming infected by *T. gondii* or that the height of the students and the probability of infection are affected by a third factor. In this case, the suspicious joint factor that simultaneously affects the height of the students and the probability of infection is the level of testosterone.

(compared to allele y) in the presence of alleles X or x causes an increase in body weight by a further 10%. Imagine that the presence of alleles X (and not x) in the presence of allele y causes a decrease in weight by 50% (i.e. the weights of animals with combinations XY, xY, Xy and xy are 120, 110, 50 and 100, respectively). The strength of genetic interactions is evidently greater than the influence of the main effects of the genes themselves. Nonetheless, if the genotype Xy occurs very rarely in the studied populations, the relevant statistical test will seem to yield quite the opposite result. The direct effects (i.e. the main effects) of both genes will be statistically significant (if we have a sufficiently large set), while the effect of statistical interactions of the two genes will be insignificant. This is a result of the fact that the significance of the relevant interactions is affected not only by the strength of the relevant genetic interactions (its effect on the phenotype of the individual), but also by the number of carriers of the relevant combination of alleles in the studied sample of individuals. If there are only a few carriers of the rare combination in the population, the strength of the particular interaction will also appear small.[6]

What follows from all of this? In my opinion, the results of genetic studies to date do not permit a qualified decision on how frequently the combined effect of a large number of genes participates in the formation of the individual traits in natural populations, how complex or how simple is the character of mutual interactions amongst participating genes and thus how the presence of these interactions can affect the ability of the population to undergo natural selection.

And why not simply test it?

Regardless of what has been said, there must certainly be a way to directly test the ability of the population to respond to selection pressures, and thus to test the validity of Dawkins' (and also Darwin's) model of biological evolution in an experiment. The population of any animal with a short generation period, for example fruit flies, can be exposed to the effect of a certain selection pressure and it can be studied whether and how well the relevant population will respond evolutionarily to the particular selection pressure. These experiments have, of course, been performed many times and their results are very interesting. In most cases, the population begins to react more or less readily to the particular selection pressure. For example, if we begin to remove small individuals from the population or prevent them from reproducing, the average size of individuals in the population will gradually increase. However, with an increasing number of generations, the population will react increasingly slower and to a lesser extent to the selection pressure. Finally, the response will disappear entirely and the increase in the average size of individuals in the population will stop completely. This phenomenon is most frequently explained by stating that the genetic variability present in the population was gradually exhausted during the first few generations. The alleles determining an increase in body weight were fixed (are carried by all the individuals in the population), while the alleles determining a reduction in body weight were removed from the population. While, at the beginning of the experiment, we selected the formerly present gene variants and thus selection proceeded rapidly, after some time this source of variability dried up and we had to select from the newly created mutations. The process of formation of new mutations is slow and thus the response of the population to selection pressure also became extremely slow.

As it frequently happens, the nice simple explanation of a phenomenon has one small defect – it is almost certainly wrong. We can easily convince ourselves of this, for example, by setting selection in the opposite direction in a population that has ceased to respond to our selection pressure, in this case by selection in favour of small individuals. If the population were uniform in the relevant genes, it should not respond to such a selection pressure. However, this is not the case. In actual fact, the population begins to react very readily to the selection pressure and evolution towards smaller body weight progresses at approximately the same speed as it progressed towards increasing body weight at the beginning of the experiment. However, it is perhaps even more interesting that, if selection is simply terminated and no selection is made in favour of either small or large individuals, the average phenotype of individuals in the population begins to return to the original value. This phenomenon is called genetic homeostasis.[7] It has most frequently been explained by stating that, in selecting in favour of the allele affecting body weight, we simultaneously helped some alleles of genes that occurred in chromosomes in their immediate vicinity, i.e. were genetically bonded with them. Unless recombination occurs directly in the section of the chromosome separating two genes, the fates of the alleles in the given part of the chromosome are mutually interconnected. In our experiment, some of the alleles of the genes located close to the alleles increasing body weight have a detrimental effect on the biological fitness of their carrier. When our selection led to an increase in the proportion of alleles increasing the body weight of their carriers in the population, the proportion of alleles decreasing their biological fitness also increased. As soon as we interrupted selection in favour of large individuals, natural selection directed against individuals with low biological fitness came into full force and was thus directed against these alleles. As these alleles did not change their position on the chromosome and remained genetically bonded with the alleles of the genes affecting body weight, the decrease in the proportion of alleles reducing biological fitness simultaneously led to a decrease in the proportion of alleles increasing body weight. Thus the average size of the individuals gradually returned to the original value.

Genetic homeostasis and the phenomenon of stopping of the response of the population to selective pressure can, however, be explained in a different way, specifically in a way that does not assume physical coupling of the fates of the alleles because of their close positions on a chromosome. If the formation of the individual traits is controlled by a large number of genes and their mutual interactions, the development of the genetic composition of the population is determined by the laws that can be described using the apparatus of game theory. The individual variants of the genes remain in the population because they determine evolutionarily stable strategy as we described it using the model of the dove and the hawk in Chapter 7. Greater selection pressure can shift the proportion of doves and hawks in the population from the original equilibrium values. For example, if we consistently remove hawks from the population, the proportion of doves will gradually increase. However, the fewer hawks there are in the population, the greater the advantage that this strategy brings its carrier. From a certain moment, this advantage will balance out the reduction in hawks caused by our activities. From that moment, the ratio of hawks and doves will stop responding to our intervention and will equilibrate at a new value. As soon as we stop removing hawks from the population (eliminating one of the disadvantages of "being a hawk"), the mutual proportions of the two strategies begins to return to the initial equilibrium value. This applies in sexually reproducing organisms for all, or almost all, genes. As we mentioned in Chapter 7, in sexually reproducing organisms in which the individual alleles meet (similar to hawks and doves) in the genotypes of the new offspring, there will be very few alleles whose effect on the phenotype and on biological fitness would not depend on their instantaneous proportion in the population. This means that most of the variability present in the population is maintained because of the fact that it represents an evolutionarily stable strategy. If we exert selection pressure on such a population, it will deviate from the former equilibrium and thus the usual phenotype of its members will gradually change. When the selection pressure is removed, the frequency of the individual alleles and thus also the phenotype of the organisms returns to the original value. Understandably, in experiments where small populations are usually studied (consisting of tens rather than tens of thousands of individuals), the return to the original state need not be perfect. Some alleles can completely disappear from the small population as a consequence of strong selection pressure or chance. In this case, they cannot return from anywhere after removal of the selection pressure and a new equilibrium is established in the population – the individuals retain, at least partially, the new phenotype.

What happens when a completely new mutation appears in a population? If this is a mutation that is disadvantageous under all conditions, it is very rapidly removed from the population. On the other hand, if this mutation is advantageous for most members of the population, it will begin to spread. If, in itself, it represents an evolutionarily stable strategy, then it can even force out all the other variants of the gene and become fixed in the population. However, this possibility apparently occurs only very rarely. Most of the traits of living organisms are quantitative in nature and are maintained at a certain optimal value, similar to the ratio of hawks and doves in the population, through the combination of counteracting frequency-dependent selection pressures. If a new mutation shifts a certain property in one direction, this produces an immediate selection pressure against all the alleles that affect this property in the same way. As a consequence, the carriers of the new mutation are, on average, at no advantage; biological fitness will increase for some and decrease for others. Thus, a large fraction of the new mutations will apparently end up as part of an extremely complicated network of mutual interactions amongst the great variety of alleles of all the possible genes. They cannot become fixed but can also not be completely eliminated from the population. The more complicated this network becomes, the greater the probability that further newly formed alleles will be caught up in it. The population and the entire species should thus gradually become more and more resistant to the action of the selection pressure. The species should react to selection pressure by a shift in phenotype; however, its ability to respond to selection pressure should be only temporary and, after termination of the particular selection pressure, the phenotype of the members of the population or species should return to the original state. And this is precisely what we usually observe in our experiments.[8]

Darwin's "sweet" secret

Basically, breeders had just the same experience long before the establishment of genetics and the results of their breeding activities were repeatedly used in the past as an argument against the validity of Darwin's theory of evolution. Darwin did not forget to emphasize in his famous book that sufficiently strong and targeted selection pressure can mould a race of dogs or pigeons to practically any form. What he did not emphasize was the fact that this almost always occurs at the expense of the viability or fertility of the members of this race. In a historically short period of time, we can breed a miniature pinscher or an Afghan hound, but we must face the fact that for most keepers of these breeds visits to the veterinarian will almost become part of their daily routine.

This is also true of most decorative breeds of practically any species of animal. It is thus not surprising that, when the populations of pure-bred animals are left to their fate, the phenotype of the members of the given population return to the original phenotype of their wild predecessors within a few generations. This is a different phenomenon than the return of the phenotype of an original wild form in the case of crosses between two different races. In crosses, the almost immediate return to the original phenotype is caused by the breakdown of the unique combination of alleles (responsible for the appearance of the members of the individual races) as a consequence of recombination and segregation of alleles. In the members of the same race, there is a gradual return to the original wild phenotype as a consequence of the action of natural selection which, during a few subsequent generations, removes from the population individuals with reduced viability and fertility, i.e. with the phenotype of the race bred by humans.

Summary and incitement

To summarize: Contemporary genetics cannot unambiguously answer the question of how many different genes are usually responsible for a single phenotypic trait and especially the importance of the genes themselves and of gene interactions for the development of traits. However, the information gained to date in genetics and selection experiments does not exclude the possibility that most traits are fundamentally affected by interactions amongst a large number of genes and thus even the theory proposed by the model of inter-allelic selection (i.e. the selfish gene theory) cannot explain the evolution of adaptive traits in sexually reproducing organisms. Similarly, long experience with the low biological fitness of improved races of animals and their spontaneous gradual return to the phenotype of their wild predecessor shows that the mechanism proposed by Darwin for biological evolution could hardly function in sexually reproducing species. The next chapter is probably the most important chapter in the whole book. In it, I will attempt to propose a way in which it is possible to bring the conclusions of the previous two chapters, i.e. the statement that adaptive traits cannot evolve in sexually reproducing organisms through the mechanisms of either Darwinist or Dawkinist evolution, into accord with the completely apparent fact that the evolution of these traits nonetheless obviously occurs in sexually reproducing species. (I don't want to advise you, but now you should exclaim suspiciously: "Well, we would certainly like to know how!")

Footnotes

1. See also the note 3 in chapter 9.
2. The polygenetic architecture of most traits is discussed, e.g. in *Human Mutations* 7: 283–293, 1996, *Evolution* 50: 967–976, 1996.
3. The relationship of genes with weak and strong effects is discussed, e.g., in *Hereditas* 125: 1–9, 1996, *Nature* 258: 665–668, 1975.
4. The accumulation of modifier genes that are capable of masking the effect of the original mutation is described in Fisher's famous (and very interesting) book *The Genetical Theory of Natural Selection*. Dover Publications, New York (1958).
5. The role of epistasis in the genetic architecture of traits and in evolution is discussed, e.g., in *Genetical Research* 86: 89–95, 2005, *Evolution* 50: 1042–1051, 1996, *Evolution* 59: 2333–2342, 2005, *Genetical Research* 74: 291–302, 1999, *Nature* 435: 95–98, 2005, *Theoretical Population Biology* 67: 141–160, 2005, *Evolution* 61: 1017–1032, 2007. Molecular data related to the subject are shown, e.g. in *PLoS Biology* 5: 922-931, 2007.
6. The difference between statistical and genetic interactions and the consequences of this difference for the genetics of quantitative traits are discussed, e.g., in *Theoretical Population Biology* 59: 61–86, 2001 and in *Evolution* 57: 706–716, 2003.
7. The response of the population to selection pressure is discussed, e.g. in the book Sheppard, P.M., *Natural Selection and Heredity*. Hutchinson University Library (London), 1958.
8. Genetic homeostasis is described and discussed by I.M. Lerner in *The Genetic Basis of Selection*. Willey (New York), 1958. A modern review of relevant data including results of long-term *in vitro* evolution experiments and data of many paleontological studies was published in *Paleobiology* 31: 133–145, 2005 by "Dream Team" of nine leading paleontologists and geneticists.

CHAPTER 11 And where does biological evolution come from, then?

It follows from the previous two chapters that biological evolution by processes assumed by both Darwin and Dawkins is basically not possible in sexually reproducing organisms. Classical Darwinist evolution occurring through the mechanism of individual selection cannot function here because the genotype, and thus biological fitness, is not inherited. Dawkins-Hamiltonian evolution, based on interallelic selection, can also not occur because, as a consequence of genetic interactions on various genetic backgrounds (i.e. in combination with the other alleles of other genes), the same alleles have a different effect on the phenotype of the individual and thus also on the biological fitness of its carrier and, in addition, the fate of the individual alleles is not determined by their direct impact on biological fitness, but rather their manifestation from the standpoint of evolutionary stability.

Elastic world

In the previous chapter, we stated that a species should respond to sufficiently strong selection pressure like elastic: It should initially give way very readily but, the further away it gets from its original phenotype, the less and less readily will it respond to the same pressure until, from a certain instant, it stops responding completely. After the end of the selection pressure, it should return to its original state, to its original phenotype.

However, evolution can hardly occur in such an elastic world, i.e. evolution in the sense of gradual changes in the traits and appearance of organisms accompanied by accumulation of adaptive traits, increasing the ability of organisms to utilize the resources present in their environment and to resist the pressure from enemies. All the knowledge gained over the past 150 years concurrently indicates that evolution is a real process in the history of the Earth. It can, of course, be pointed out that evolution in sexually reproducing organisms can be driven by other processes than natural selection. For example, in the introductory part of the book, we discussed genetic drift, genetic hitchhiking and evolutionary drives and sorting from the standpoint of stability. All these processes can occur in both asexual organisms and in sexual organisms. However, none of them can satisfactorily explain the formation and development of adaptive traits. Even if a large number of adaptive traits function at an intracellular level and could thus have been developed in single-cell organisms prior to the development of sexuality, a considerable portion of adaptive traits developed only in multicellular organisms and thus definitely in sexually reproducing organisms. No, if we want to manage without the intervention of supernatural forces (and, in science, we should at least attempt to do this), then we need to explain the formation and

Box 11.1 Supernatural forces and science

Science cannot decide on whether supernatural forces exist or not. If there were a God who did not have to obey the natural laws of our world, then he could arrange for the experiments of scientists to have any results whatsoever and they could thus never either discover or exclude his existence. The explanation of a certain natural phenomenon based on the assumption of the action of supernatural forces is thus not scientific and is bad because it is necessarily erroneous or because science does not recognize the existence of God. Science cannot decide whether this is erroneous or not. It is unscientific because no consequences follow from it that scientists could test and thus potentially falsify. It is simply not possible to test the lack of correctness of a supernatural explanation and thus, in science, we must always attempt to explain the observed phenomenon by natural means – by processes not including the action of supernatural forces. It is quite possible that we will never be able to explain some phenomena by natural means; but this does not change matters. If evolution or God gave us reasoning, we must try as honestly as possible to use it to understand our world.

existence of adaptive traits in contemporary organisms by natural selection in all types of organisms without regard to their means of reproduction.

The riddle and its solution – It's elementary, my dear Watson

And now I would like to present a nice circus trick that I learned from the authors of detective stories. Dearest readers, at this moment, I have already presented you with all the information required to successfully resolve the described riddle. It is sufficient to put the pieces of the puzzle together properly and you can find the answer to the question of how and when natural selection in sexually reproducing organisms can lead to the development of adaptive traits. It is so elementary, my dear Watson (no, it is not related to the structure of DNA, I meant a different Dr. Watson). It is enough to simply know which two pieces of the puzzle should be fitted together first.

Well, do you know the solution to the riddle? I know, it's not entirely elementary. Perhaps it would help to first go and make a cup of tea or take the dog for a walk (he was whining strangely some time ago) and to think about it for a while. If you did not skip the chapter on speciation, then you could figure it out without any hints.

Shall we continue? Tea made, dog walked? So, a hint. Actually, I already gave it to you, but it was sneakily camouflaged. I asked how and WHEN evolution of adaptive traits could occur in sexually reproducing organisms. It indirectly follows from this that situations probably occur in which the evolution of adaptive traits can occur even in sexually reproducing organisms through the action of classical Darwinist selection (and just as easily through the action of Dawkins-Hamiltonian interallelic selection). Getting warmer? Then careful; in the following paragraph I will tell you the solution to the riddle, so this is the last chance to solve it by yourself. (The dog is whining again and the flowers in the living room need watering…)

Okay, then let's get on with it. During the existence of any species, a period can (but need not) occur when this species is just as susceptible to the action of natural selection as a species that reproduces asexually. This period is speciation and especially the period immediately following speciation. This applies to all species that evolved by splitting off of a small number of individuals from the population of the parent species (i.e., for example, species that were formed by certain forms of sympatric speciation, for example polyploidy speciation, and also species that were formed by the mechanism of peripatric speciation – see Chapter 6). If the new species is formed from a small number of individuals, then it takes with it only a small part of the genetic variability of the parent species. The members of the forming species are thus genetically if not completely identical then at least very similar. This genetic similarity can in no way be altered by segregation and recombination of chromosomes (see Chapter 3) – genetically identical individuals can segregate and recombine as much as they want – still they remain genetically identical. (It is like if we were to extremely carefully and lengthily shuffle a packet of cards consisting of only aces of hearts.) As a consequence, a new allele of any of the genes in each generation and in each individual is necessarily in the company of the same alleles, i.e. in the same or very similar genotype. The effects of alleles on the phenotype of an individual and thus on the genetic fitness of its carriers does not change much under these circumstances, and this subsequently allows functioning of classical Darwinist selection.

And that's it. According to the results of this model, which we will now call the **theory of frozen plasticity**, sexually reproducing species exert evolutionary plasticity only in the first stage of their existence, specifically only until such time as sufficient genetically determined variability accumulates in their gene pool.[1] The consequence of accumulation of this variability is the formation of a complex network of mutually interconnected and mutually determining evolutionarily stable strategies capable of maintaining the species for a long time in the state of genetic homeostasis, i.e. in an evolutionarily frozen condition. In the history of species that formed from small founding populations, two periods alternate: The relatively short **period of evolutionary plasticity**, in which the species can change its properties in response to selection pressures in the environment and thus form new, adaptive body organs and new adaptive patterns of behaviour, and then a very long **period of frozen plasticity**, during which the species can

Box 11.2 Extinction of species

At the present time, the extinction of a species is considered to be the instant when the last representative of the particular species dies. "Pseudo-extinction", i.e. the gradual change of one species into a different species, is not considered to constitute extinction; some palaeontologists believe in the existence of this process that, according to the theory of frozen plasticity, should occur only in asexually reproducing species. Palaeontological data indicate that extinction is the unavoidable fate of every species. The average time of survival of species differs for the individual taxons. For example, the period of survival of the average mammal species varies around 5 million years, while the period of survival of sea snails and clams is about 10–20 million years. Species become extinct, either as a consequence of sudden catastrophic events, for example the impact of asteroids or the cores of comets on the Earth, or gradually, as if there were no external cause at all. Some facts indicate that the commonest cause of gradual (called background) extinction consists in pandemics caused by a parasite, probably most frequently a virus, see also Footnote 3 in Chapter 16.

respond to selection pressures only temporarily and in a very limited manner and thus only passively waits for the moment when a substantial change in its environment leads to its extinction.

Brave plastic world

In order to be more precise, the first stage could be divided into two time periods. In the period immediately following separation of part of the population of a new species from the remainder of the population by a sufficiently strong isolation barrier, the population of the new species has **few members**. In a small population, any kind of selection has very low effectiveness as mostly genetic drift (chance) decides the fate of the individual alleles here. Thus, during this first period, the alleles with a selection advantage cannot spread sufficiently effectively in the population, although the high heritability of the individual traits and the high heritability of biological fitness otherwise create favourable conditions for the functioning of natural selection.

In the first period, there is a further decrease in the genetic variability of the population, where this decrease is related to the most important components, the alleles, which are usually maintained in the population through frequency-dependent selection. As was already mentioned in Chapter 6, some formerly common alleles could have completely disappeared in small populations and, on the other hand, formerly very rare alleles could have become frequent. Thus, if a new species takes with it part of the mutually dependent genetic variability, in a period when the population is small, this network cannot function because of the limited effectiveness of selection (see the chapter devoted to genetic drift), the particular selection pressures weaken and a great many alleles disappear from the population through the effect of genetic drift.

The most important event of the subsequent second period is a substantial **increase in the size of the population**.[2] Understandably, not every newly formed species goes through the period of population growth. However, we will never learn of species that did not substantially increase in population after their formation. It also holds in nature that the winners write history. A species whose population remained small after its formation is exposed to a much greater risk of extinction as a consequence of local environmental changes. Even if this did not happen for a long time, its members will still probably not get into the palaeontological records. Only a negligible number of individuals of relatively successful species are preserved in the form of fossils. It is basically impossible to find a fossil of a species with a small population. Most speciation certainly ends with complete failure – the new species dies out soon after its formation, or merges back

Box 11.3 Heredity and heritability of traits:

Heredity is the ability of traits (certain properties that adopt at least two forms in the population) to be inherited from parents by their offspring. Heritability of traits expresses the degree to which a certain trait is inherited from parents by progeny. The heritability of traits can be expressed as the fraction of genetically determined variability in the given trait in the total (i.e. determined genetically and through the effect of the external environment) variability of this trait.

into the original species. From the standpoint of the history of evolution, these unsuccessful attempts do not count – history is written by the successful.

An increase in the number of members of a new species (for example, on a newly colonized territory) is an instantaneous matter on an evolutionary scale. During a few dozen generations, the population of a new species can increase to numbers comparable with the number of the original species. I should point out that, at the time when the population is increasing in number, the effectiveness of natural selection is even less than in the period of stagnation of the population, as even the weakest, who would submit to intra-species competition in a constant population, live to adulthood. After the limits of sustainability in the particular environment are achieved or, to be more precise, after a certain limiting density is achieved (see Chapter 15) the growth in the population will end and the size of the population will begin to fluctuate around this value for a long period of time. And the species begins to evolutionarily freeze. However, the process of evolutionary freezing of the species occurs on much longer time scales than the increase in the size of the population. Thus, for quite a long time, the species exists under conditions that are completely ideal from the standpoint of natural selection. It forms a large, genetically rather homogeneous population, in which the effectiveness of selection is very high in comparison with the effectiveness of genetic drift. Simultaneously, the substantial genetic uniformity of the population ensures that the heritability of phenotypic traits, and thus of biological fitness, will be high. In this second period, the evolution of adaptive traits apparently occurs most rapidly; the species best adapts to the conditions of its environment through the gradual accumulation of mutations.

Brave frozen world

However, over time, new mutated alleles are gradually formed in the population, whose effect on biological fitness depends on their frequency in the gene pool of the population; specifically, their effect becomes smaller as their proportion in the gene pool increases. These mutations are then maintained over time in the population by frequency-dependent selection (see the example on plundering orchids in Chapter 7). Of course, other types of mutations are also formed in the population; however, these are regularly removed from the population by selection or genetic drift, or become completely fixed in the population. Thus, they need not momentarily interest us from the standpoint of evolutionary freezing of the population. The genetic variability of the population increases with each allele that is caught in the population and is further maintained by frequency-dependent selection and thus the probability increases that a subsequently mutated allele will end up amongst alleles whose effect on the phenotype and biological fitness is affected or even determined by the alleles borne by the relevant individual. Thus, the species enters its stabilization phase at an increasing speed, its evolutionary plasticity decreases and it becomes frozen at an increasing rate.

From the standpoint of resistance to ecological effects, e.g., short-term fluctuations in the climate, evolutionary freezing of a species is not as disadvantageous as would seem at first sight.

The phenotype and composition of the gene pool of the population of an evolutionarily plastic species continuously follows changes in the environment. As a consequence of temporary, sufficiently substantial changes in the local natural conditions, selection can remove from the population, or even the species, those alleles that would be very useful when the natural conditions return to normal. A species whose evolutionary plasticity is frozen adapts only partially to temporarily altered natural conditions and only shifts the frequency of the individual alleles to a new equilibrium state. After conditions return to normal, the frequency of individual alleles, and thus the phenotype of the particular species, also very rapidly returns to normal.

It is most surprising that, in the light of the theory of frozen plasticity, sexual reproduction plays quite the opposite role to that which is automatically assigned to it by most evolutionary biologists. The development of sexual reproduction is one of the greatest mysteries of evolution. Sexual reproduction is, taken overall, a pleasant affair; however, it entails a number of unpleasant consequences for the individual. Paul, don't trample on my keyboard, can't you see I'm working? Jane, get off my head, do you want daddy to have a stiff neck again? There are a number of good reasons why organisms should reproduce asexually and why progeny of asexually reproducing mutants should predominate in the population over time. One of these reasons, for example, is the fact that an asexually reproducing mutant female, who does not waste half her reproductive capacity in the production of males, should theoretically reproduce at twice the rate of her sexually reproducing competitors. Over time, evolutionary biologists have thought up a vast number of hypotheses to explain the general existence of sexual reproduction in nature.[3] Amongst the more important hypotheses are models based on the idea that sexual reproduction reduces the immediate competitiveness of the individuals, but increases evolutionary potential (capacity) of sexual species in comparison with asexually reproducing species. For example, some of them assume that sexual reproduction permits the long-term existence of a diploid genome, where the existence of two copies of each gene allows the organism to perform evolutionary experiments with its genetic material, specifically with one of the two copies of the gene, and thus to create (by the method of trial and error) new genes coding new proteins. Other hypotheses assume that sexual reproduction facilitates the long-term existence of genetic variability within the species and thus provides the evolutionarily substantial source of variability required for a rapid evolutionary response to newly emerging selection pressures. On the contrary, the theory of frozen plasticity indicates that the main advantage provided by sexual reproduction consists in a substantial reduction in the evolutionary ability of most species. As a consequence of sexual reproduction, most species are evolutionarily passive throughout much of their existence and cannot opportunistically (i.e. without regard to future negative consequences) respond to temporary short-term changes in external conditions. However, in this way, they retain their adaptation to the usual conditions of their environment over long periods of time. But I would not want to be unjust to evolutionary biologists – for example, in his book "Sex and Evolution", published in the 1970s, George C. Williams pointed out the potential effect of evolutionary passivity in sexually reproducing species. The fact that his ideas did not receive an appropriate reception is not a mystery of evolutionary biology, but rather a mystery of sociology of science.

Frozen species are better able to resist temporary, irregularly occurring changes in the environment, but they cannot adapt to new conditions. Consequently, in their range of occurrence, they are found in a decreasing number of places (and tend to form smaller populations there) until their frozen phenotype becomes completely "obsolete" and they die out.[4]

Summary and incitement

The main message of this chapter and, basically, of the whole book should be that the inability of most sexually reproducing species to respond evolutionarily to selection pressure can be harmonized with our experience that the evolution of sexually reproducing species quite obviously occurs. If a new species is formed by splitting off of a small part of the population from the original parent population, then it will take with it from the parent population only a small part of the genetically determined variability and, as a consequence of the resultant genetic similarity or even identity of the individuals in the population, can be subject to Darwinian evolution. However, after a longer period of time, the gene pool of the new species accumulates new variability and the species again evolutionarily freezes. Surprisingly, evolutionary passivity of sexually reproducing species can be responsible for their evolutionary success. From a long-term standpoint, it allows them to better maintain the best (most useful) phenotype in an environment where natural conditions irregularly fluctuate for short periods of time. In the next chapter, we will see whether the conclusions of the theory of frozen plasticity agree with the evidence that palaeontologists have gained from the study of fossils.

Footnotes

1. I published the theory of frozen plasticity in *Rivista di Biologia–Biology Forum* 91: 291–304, 1998; however, I am not sure whether I am its author. It "came together in my head" when I read the creationalist book *"Darwin on Trial"* (E.P. Johnson, 1991, DC: Regnery Gateway Publishing Company, Washington, DC, 1991) and I consider the combination of the genocentric Dawkins-Hamiltonian model of evolution with the theory of evolutionary strategies of Maynard-Smith to be its main idea basis and also my main contribution to the theory. I collected genetic documentation for the validity of this theory (the results of selection experiments, analysis of the genetic architecture of quantitative traits) only subsequently and I discovered much later that Eldredge and Gould originally proposed an, in principle, a very similar explanation for their palaeontological data in *Models in Paleontology* 5: 82–83, 1972. I sent the first version of the manuscript to *BioEssays* in 1997 (it was immediately returned without a formal review) and also to J.S. Gould (he did not answer but, considering the number of letters he probably got from the readers of his books, I wasn't really surprised) and to Maynard Smith (he answered in a very nice, handwritten letter that he did not agree with my conclusions (and the basic starting point – the ineffectiveness of selection in sexually reproducing species)).
2. In the discussion following my lecture devoted to frozen plasticity (presented on October 16, 1997 in the framework of the series of lectures "Thursdays in Viničná" (Prague, Czech Republic)), Daniel Frynta pointed out the existence of two stages in the period of evolutionary plasticity and the importance for the development of a species of a rapid increase in the number of individuals in the still genetically homogeneous population. This is very important as the original model by Eldredge and Gould (based on ideas of I.M. Lerner and E.

Mayr) predicts that the evolution of new species proceeds in small populations rather than in large, genetically homogeneous populations. The major difference between the model of depleted genetic polymorphism and both Tepleton's model of genetic transilience (*Genetics* 94: 1011-1038, 1980) and Carson's founder-flush model (The population flush and its genetic consequences. In: Lewontin RC, Ed. *Population Biology and Evolution*. Syracuse, N.Y.: Syracuse University Press, 1968) is that the former model suggests that a decreased, rather than an increased, level of variability exists in the founder population and is responsible for the increased evolutionary plasticity of the new species. I do not know which of these models (if any) is correct (actually, all of them could be). However, any one of them is more or less compatible with the theory of frozen plasticity. The main advantage of the depleted polymorphism model is its simplicity - it can be easily understood by nonprofessional readers.

3. Of books on this subject, I can recommend especially G.C. Williams, *Sex and Evolution*. Princeton University Press (Princeton), 1975 and J. Maynard Smith, *The Evolution of Sex*. Cambridge University Press (Cambridge), 1978.
4. It has actually been observed that evolutionarily older species, i.e. species that branched out sooner within a certain evolutionary branch, tend to have large areas of occurrence, but occupy fewer sites in these areas, see *Evolution* 58: 2622–2633, 2004. As David Storch pointed out to me, the lower plasticity of older species could also explain why there are enormous differences in the numbers of individuals in different species. In the framework of each developmental line, we usually find only a few species with many members (which could be young, still plastic species) and simultaneously a large number of more or less rare species (old frozen species gradually becoming extinct). *Ecology* 29: 254–283, 1948; Gaston K.J. (1994), *Rarity*. Chapman and Hall, London; *Trends in Ecology and Evolution* 11: 197–201, 1996, *Ecology Letters* 10: 995–1015, 2007.

CHAPTER 12 And what does palaeontological data have to say?

It's all very well that the individual parts of the theory fit together, but this is only to be expected of a theory. It could hardly be expected that a new theory would contain obviously logical errors right from the start. For this purpose, the scientists have the above mentioned Occam's broom, to sweep the contradictory data under the carpet, and also the previously unmentioned Occam's iron to iron out all the ugly seams and folds, ends of threads or even holes in the theory. The result is the nice-looking, well-fitting coat of the new theory, nicely explaining all the questions that the author presented to his listeners at the beginning. This is caused, amongst other things, by the fact that a good author first forms his model and only then does he write the introductory chapters, in which he brings up the questions to which his model provides a satisfactory answer, as I already mentioned in Chapter 2. In any case, the internal lack of contradiction in the theory is only a very poor measure of its objective correctness. A good scientific theory can be distinguished from a bad theory in that it yields the greatest number of practical results whose validity can be subsequently tested. It is encouraging that the theory of frozen plasticity complies with quite stringent criteria exactly in this respect. As will be shown below, this follows from a number of practical consequences that can become and, in some cases, have already become the subject of scientific testing.

On missing links and evolutionary leaps

Palaeontology – the study of fossils – can probably provide us with the most illustrative examples of the validity of the theory of frozen plasticity. Both the Darwinist and Neodarwinist theories of evolution assume constant evolutionary plasticity of species. Consequently, they predict that the traits of representatives of individual species, and thus the shapes and sizes of the relevant fossils, change over the entire existence of the species. In contrast, the theory of frozen plasticity predicts that species change, i.e. differ from their predecessors, only in a short period of time following their formation and subsequently, for a long period of the order of several million years of their existence, remain invariable or undergo only short, temporary changes in dependence on the variability of the conditions in their environment. Thus, the theory of frozen plasticity predicts that changes in the appearance and traits of species should be bound in time only to the period of speciation, i.e. a time period of closer to tens of thousands of years, or even less. As this period covers only a tiny fraction of the life of the species, there is only a very small chance that a fossil will be preserved from this period. The lack of completeness of the palaeontological record is reflected, e.g., in the fact that only one specimen has been found for half of the 350 described species of dinosaurs, or that we occasionally catch in the sea an animal that belongs in a taxon whose last members were thought by palaeontologists to have become extinct tens or even hundreds of millions of years ago (for example, the Coelacanthiformes fish *Latimeria chalumnae*)[1]. The number of preserved fossils of marine vertebrates is understandably much larger than the number of fossils of terrestrial vertebrates. This is a result both of the more favourable conditions for fossilization in marine sediments and also the fact that their populations were always much larger. The probability that the body of an individual will undergo fossilization after death is extremely small in the normal environment and the vast majority of individuals disappear after death in the stomachs of other organisms, or simply decay. One of the consequences of the incompleteness of this palaeontological material could be that, in accordance with the theory of frozen plasticity, new species should appear in the palaeontological record **suddenly and in their final form.**

In contrast, the Darwinist and Neodarwinist models assumed that the new species should initially have an appearance similar to that of the species from which it was formed and should then differ from it more and more over time. If we look in the textbooks of evolutionary biology and palaeontology published practically up to the 1980s, the theory of frozen plasticity does not look too probable. The gradual transformation of the species and the occurrence of transition forms between the individual species was frequently given as one of the best proofs of the correctness of Darwin's model of evolution. It is true that transitional forms were found in palaeontological records somewhat less than biologists might have wanted. However, evolutionary biology was more or less satisfied with the explanation proposed by Charles Darwin that the palaeontological record is incomplete. Palaeontologists had nonetheless known for quite some time that, at least for some groups of organisms (especially the above mentioned marine invertebrates), the palaeontological record is so good and complete that the lack of intermediate links or the absence of changes in the body structures of organisms during the existence of the species requires a different explanation. However, the non-existence of whatever usually has the character of a negative result and scientists then very easily succumb to the natural tendency to place the results of such studies in the bottommost drawer.

Box 12.1 Negative and positive results

A scientific study is performed in an attempt to support or negate the validity of a certain hypothesis. In the optimum case, the study has the character of a cross experiment – a certain result would support the studied hypothesis while the opposite result would throw it into doubt. In this case, the study provides only one of the answers, yes or no. However, very frequently, our studies have an asymmetric output. For example, one result supports our theory, but the opposite result does not mean anything at all. In this case, the study can yield the answer yes (no) or do not know. In the second case, this is called a negative result. You would probably like a specific example; here is one: If we do not manage to find a transition link in the palaeontological record, then this can mean that that link never existed, but it can also mean that a fossil has not been preserved, or that we have simply not found it yet. However, if we find the intermediate link, our hypothesis will be supported.

In addition, the editors and reviewers of journals will also unselfishly help him come to this conclusion. Publish the results that demonstrated the existence of a change – of course, every well-supported document of the validity of a generally accepted theory is certainly welcome and deserves to be published in a prestigious journal. On the other hand, it is a waste of paper to publish a study that did not demonstrate any statistically significant change, that, in fact, did not demonstrate anything at all. "My dear colleague, that is most unfortunate for you, but that's science, you will undoubtedly be more successful next time. Try studying three other species; we all know that species must have changed over time." And so it happened that evolutionary biologists did not realize for a long time that modern palaeontological data could be contradictory to their beloved theory and palaeontologists did not realize that the non-existence of a certain phenomenon could, in actual fact, be the most fundamental result that they obtained during their scientific career, a result that is capable of shaking the very foundations of the 150-year-old theory of evolution. And so it was that the princess together with the entire kingdom slept their deep and long sleep and waited for the prince who would battle his way through the thorns of scientific self-censure and disfavour of reviewers and wake them up with his kiss.

Wakening of the Sleeping Beauty and theory of punctuated equilibria

Basically, there were two ways of awakening the princess. As I am rather malicious, I would prefer the scenario in which the main role would be played by a curious and hard-working creationist. If a creationist were to decide to test the basic pillar of Darwin's model of evolution on sufficiently extensive palaeontological material, i.e. to test whether species change over their existence, he would certainly discover (without regard as to whether this is true or not) that they are invariable throughout the period of their existence. And if he published his results with sufficient humbug, scientists would be forced to react in some way. The nonexistence of change would cease to be a "non-phenomenon" and would become a phenomenon worthy of scientific study. As the reader might have noticed, I do not usually have excessive illusions about the unbiased opinions and objectivity of science and scientists. Simultaneously, however, I am of the opinion that, as soon as an individual aspect becomes an object of the interest and study of scientists, science as a whole is capable of finding the right answer to the particular question. If palaeontologists were, in time, to confirm the original result on the invariability of the species, it would be up to evolutionary biologists to find an explanation for this phenomenon, to propose a suitable mechanism. For example, the theory of frozen plasticity.

However, unfortunately, the scenario with the curious and hard-working creationist has not occurred. Probably because these individuals occur in our world with about the same frequency as sleeping princesses. It's no fun to shut yourself in a study for several years and gradually measure thousands or tens of thousands of fossils, in addition with the risk that nature (or God) will finally provide us with quite the opposite answer to what we originally expected, or would like, to obtain. In the end, the princess was awakened by a scientist, to be more exact two scientists. These were the palaeontologists Niels Eldredge and the palaeontologist and simultaneously evolutionary biologist, historian of science and writer of popular science articles, Stephen Jay Gould. The discovery was originally published by Eldredge, however, it was probably the all-round talent of Gould that permitted recognition of the real importance of the phenomenon-nonphenomenon of the nonexistence of changes in the properties of organisms during their existence. The manner in which the pair of young authors overcame the vigilance of editors and reviewers is also quite characteristic. Eldredge and Gould did not attempt to overcome this obstacle (at least as far as I know), they simply went around it. Rather than in an established palaeontological journal, they preferred to publish their fundamental article in the unreviewed proceedings of a conference. In any case, as a survey of famous scientists indicates, this approach is basically the rule in the case of fundamental scientific discoveries.[2] A truly original and simultaneously important work has little chance of being accepted for publication in a good or at least average journal. In the role of reviewers, we are probably not always capable of understanding the results obtained in all their detail or of objectively evaluating the correctness and reliability of the methods employed. However, what we can almost always recognize is the originality of the results. And as scientists, even in the role of reviewers, we must primarily think statistically. What is more probable – that the editor has sent me the work of a genius to be reviewed, a work that is capable of causing a revolution in my field? Or that they sent me the publication of some joker who skilfully twisted or even thought up the experimental data? Hmm, there are far more jokers and falsifiers of data than geniuses. In addition, if I reject the manuscript, even unjustifiably, more or less nothing will happen to me, an anonymous reviewer. What is more, I will show the editor how critical I can be of other people's results, and thus, probably, especially of my own results. However, if I recommend a doubtful manuscript for publication, then I can get myself into a very embarrassing situation, at the very least in the eyes of members of the editorial board of the journal. So why hesitate? Thumbs down.[3]

Some of the works published in journals with less strict review processes are surprisingly not ignored and, in time, find their way to the relevant public. This was fortunately the fate of the work

Box 12.2 Model of punctuated equilibria

Eldredge and Gould called their discovery the "Model of punctuated equilibria". The name is intended to express their concept that evolution occurs as an alternation of short periods of tumultuous change followed by a long period of evolutionary calm (stasis). Thus evolution is not **gradualistic**, occurring as a slow, more or less regular change in shapes and functions and smooth transition of the older species into a new species, but rather **punctuated**, and is characterized by jerky development that occasionally occurs rapidly, but with long intervals when nothing at all happens. As I have discovered on extensive experimental material from students, punctuated evolution (in Czech punktuacionalistická evoluce) is a term that is impossible to remember or at least enunciate. Anyone who can say it rapidly three times in a row (in Czech) can consider himself to be an experienced evolutionary biologist and can, without trepidation, apply for the position of head of a department of evolutionary biology at any Czech university. (I have not managed yet, but that does not matter, because no Czech universities have a department of evolutionary biology.)

by Eldredge and Gould.[4] A number of palaeontologists recall the uncompleted, unsent or rejected manuscripts lying at the bottom of their drawers. Others calculated that, under the present conditions, no study based on following evolutionary changes in a species could yield a negative result. Either the result will agree with the traditional gradualistic model of biological evolution or will, to the contrary, agree with the new **model of punctuated equilibria** of Eldredge and Gould.

Works began to appear in professional journals like mushrooms after the rain, confirming or rejecting the new theory on independent palaeontological material. So what is the situation now, about 40 years after publication of this famous study? I am of the opinion that the new model of evolution is very well off. It makes sense to test the theory particularly on species of marine invertebrate fauna with hard shells. They occurred in shore waters in such numerous populations that their fossils can be found in adjacent layers of sediments in such large numbers that it is possible to monitor changes in their body structures during long geological periods. Punctuated evolution is encountered in the vast majority of cases in marine multicellular invertebrates (for example, clams). In contrast, amongst marine single-cell organisms with hard shells (e.g. Foraminifera), gradualistic evolution tends to predominate.[5] However, it should be borne in mind that a number of cases originally given as examples of gradualistic evolution were found, on closer inspection, to be a gradual change in the relative proportions of two morphologically different species that occurred together in the relevant area from the very beginning.

Two explanations (and how to select the worse one of the two)

What was the mechanism proposed by Eldredge and Gould for explanation of the punctuated character of the evolution of the species? Basically, they proposed two possibilities in their original work. The first possibility, in my opinion the correct one, was finally abandoned and they became more inclined towards the second, less radical one. Even young Turks age in time. According to the first hypothesis, **genetic revolution** is responsible for the evolutionary plasticity of a newly formed species, as was described by Ernst Mayr in his famous book.[6] (Those are two advantages of living a long life, the creation of a number of important works during one's lifetime and, if possible, living long enough for someone to remember our earlier, unjustly forgotten results. Ernst Mayr, who celebrated his one hundredth birthday in 2004 in relative mental comfort, necessarily managed substantially more than his similarly brilliant colleagues. Just think of it, just that cumulative citation index alone! However, nothing should be overdone. Mayr managed to live long enough for his newly rediscovered ideas to be forgotten again.)

Especially in connection with peripatric speciation, Mayr pointed out that a small population that splits off from a larger population takes only a small part of the genetic variability with it. This can subsequently lead to further dramatic changes in the genetic composition of the population that, of course, can be accompanied by the corresponding changes in the appearance of the members of the new species. (I spoke about genetic revolution on p. 87; but I did not want to go into excessive detail in the chapter on speciation, so as not to make it too easy for the reader to

Box 12.3 Cumulative citation index

The cumulative citation index is the total number of cases in which the results of a certain author published in one of his articles were cited in the scientific articles of his colleagues. In practice, mostly only citations in journals included in the Web of Knowledge database (the new form of the original Science Citation Index database) are included and auto-citations are also included – i.e. cases, where the author or one of the co-authors of an article cited this article. Both are substantively incorrect, but technically readily feasible. The older an author is, the larger the value of his cumulative citation index becomes. Thus, it is usual to assess the quality of a scientific worker or scientific team according to the number of citations achieved over a certain shorter period of time, for example, over the past 5 years. In this case, older authors with a greater number of formerly published works are, of course, also at an advantage, because their older works can also be cited in this period of time. However, in fields where works age rapidly, for example, in molecular biology, this advantage is not great. It is a matter of speculation whether this sufficiently compensates older workers for the disadvantages entailed in the obligation to sit in various commissions and councils.

solve the riddle on p. 140, specifically looking for a way to reconcile the theoretical impossibility of evolution of adaptive traits in sexually reproducing organisms with the fact that the evolution of adaptive traits undoubtedly occurs in these organisms in spite of the theory.) In large populations, the evolutionary success of alleles is measured in terms of their ability to cooperate as well as possible with the largest number of alleles of their own gene and with the alleles of other genes. In contrast, in the split-off population, in which only a small part of the originally present genetic variability remained, the alleles themselves have a greater effect on the biological fitness of their carrier. In the 1940s, Mayr of course knew nothing about evolutionarily stable strategies, but he was well aware of the importance of epistatic interactions (see Box 10.1 on p. 131) for the effectiveness of natural selection. Intuition and a talent for observation apparently told him that a genetic revolution, which could occur during some types of speciation, could play a fundamental role in the process of biological evolution.

The second mechanism that Eldredge and Gould proposed as an explanation for the punctuated character of evolution is the **substantial decrease in the size of the population** accompanying every peripatric speciation (the formation of a new species from a few founding members outside of the main area of occurrence of the parent species). In this case, Eldredge and Gould indirectly referred to the work of one of the spiritual fathers of Neodarwinism, Sewall Wright, and his **shifting balance model**. Be careful, try not to confuse the punctuated equilibrium and shifting balance theories. Wright emphasized that a temporary decrease in the size of the population, necessarily accompanied by a substantial reduction in the effectiveness of any kind of selection, can surprisingly lead to an increase in the probability of evolution of adaptive traits. Only under conditions where natural selection is temporarily less effective do mutated individuals with transitional traits have a chance of survival and of leaving a sufficient number of progeny; these individuals are not well adapted to use the ecological niches of the original species and, simultaneously, they have not yet adapted well to utilizing the future niches of the new species. As

Wright pointed out metaphorically, mutated individuals present on the slopes of hills or even deep in the valleys of an "adaptive landscape" do not have a great chance of survival in a large population (Fig. 12.1). And without going down into the valley of an adaptive landscape, it is not possible to occupy new peaks in this landscape.[7]

Wright's shifting balance model, explaining the role of a temporary decrease in the size of the population, is somewhat more complex than the brief description given here. Because of this complexity and because of the complexity of the mathematical apparatus that Wright employed in his work ("Who can bother working through those complicated integrals, come on, let's go and measure the scales of cichlids"), this model gradually fell into disfavour amongst evolutionary biologists. Perhaps this is the reason that Eldredge and Gould did not refer to it extensively and tended to emphasize more some **other advantages of small populations**. A great many of these populations are formed along the edges of the area of occurrence and each of them constitutes a separate evolutionary experiment. This substantially increases the chance that a successful new

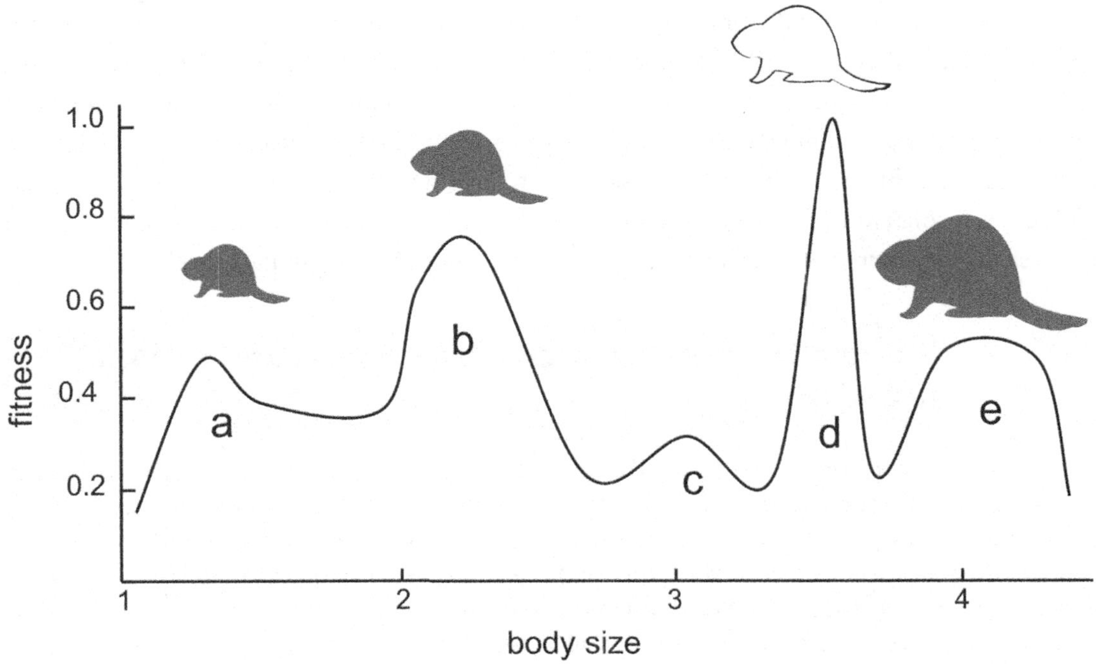

Fig. 12.1 Adaptive landscape. The scheme depicts an adaptive landscape with four adaptive peaks. The heights of the curves above the horizontal axis indicate the biological fitness of the individuals of a particular size. The four peaks of the curve correspond to the four optima, i.e. the sizes that are most advantageous from the standpoint of survival and propagation of species in the particular environment. In order for a species to be able to occupy the highest peak, some of its members (probably at peak e) would first have to go down into the valley of the adaptive landscape (become smaller as a consequence of accumulation of the relevant mutations), i.e. have a phenotype that is disadvantageous from the standpoint of biological fitness, and only then, as a consequence of further mutations, climb to a new unoccupied peak (become even smaller), i.e. gain the phenotype that is optimal in the particular adaptive landscape.

species will be formed from one of these populations. Because these populations are small, all the evolutionary changes occur more rapidly here than in large populations. Small populations arise along the edges of the area of occurrence of a large population, i.e. under conditions that are somehow extreme from the standpoint of the existence of the species (e.g. extremely cold or hot conditions for the particular population). Thus, their members are exposed to the corresponding extreme selection pressures. The populations of species formed by splitting off (peripatric speciation) occur on a limited territory. In this small territory, the conditions are the same or at least similar everywhere, in contrast to the area of occurrence of the parent species. Selection pressures acting on the members of the newly forming species are thus also the same everywhere. Isolation from the area of the parent population simultaneously ensures that no immigrants will bring foreign alleles into the gene pool, whose presence could reduce the chance of the population adapting perfectly to local conditions. Eldredge and Gould concluded that, because of the simultaneous action of all these factors, it is not surprising that a successful evolutionary experiment, the formation of a new species differing substantially from the original species, almost always takes place in a small population located out of the main area of occurrence of the parent species and not directly in the large population in the main area of occurrence. What we finally see in the palaeontological record as the practically instantaneous replacement of the old species by a new species is not, in fact, the evolutionary process of the formation of a new species at the particular place, but an ecological process, the forcing out of the old species by the new species, which was formed much earlier and somewhere else and, at the particular instant, only expanded the area of its occurrence at the expense of the occurrence of the "original" species.

Box 12.4 The relationship between a hypothesis and a model and a theory

There is not usually any difference between a hypothesis and a model in science. A model is basically our hypothesis of the nature of a phenomenon. (However, not every hypothesis need be a model; some hypotheses are not related to the nature of things, but only to the existence or nonexistence of a certain phenomenon.) In technical fields, models are intended so that study of their behaviour in cases where this is advantageous or even necessary can replace study of the behaviour of the actual, modelled object. Models are created in science so that we can test their validity and thus reject the relevant hypothesis. A **theory** is actually a more complicated hypothesis, to be more exact, it is a system of several or a great many interconnected hypotheses. The usual concept of lay persons that a hypothesis is an insufficiently verified theory certainly does not hold true.

As I mentioned above, I think Eldredge and Gould rejected their original idea entailing the key role of genetic revolution somewhat prematurely. To be more exact, S.J. Gould explicitly distanced himself from this theory in his monumental and, unfortunately, last work, "The Structure of the Evolutionary Theory", just before his death in 2002.[8] I think that this was a textbook example of methodically correct but substantively erroneous use of **Occam's razor**. Finally, we have gotten around to it. While the formerly mentioned Occam's broom and Occam's iron are intended more as a joke, Occam's razor is a useful methodological instrument named after the important mediaeval scholar, William of Ockham. His recommendation "*Entia non sunt multiplicanda, praeter necessitatem*" is currently considered as the very reasonable requirement that, of two models that are equally capable of explaining the nature of a particular phenomenon, the scientist should prefer the simpler one. Not because scientists would naively think that nature is simple. However, models (hypotheses) are primarily intended for future attempts to overturn them. And simple models (hypotheses) are easier to overturn than complicated ones.

And now, back to genetic revolutions. Eldredge and Gould concluded that, in the explanation of the punctuated character of the palaeontological record, it is sufficient to employ the generally accepted phenomenon of peripatric speciation, and thus that their model need not include the strange phenomenon of genetic revolution, of which most evolutionary biologists have never heard anyway, and the remainder mostly do not believe in it. From a methodological and tactical standpoint, this was certainly the right decision. A simple model based entirely on the action of generally known and generally accepted processes understandably has a much better chance of being accepted than a more complex model that, in addition, assumes the action of a very unusual process, whose existence is, to say the least, the subject of rejection or even collective oblivion.

However, from the substantive point of view, at least in my opinion, it was an erroneous decision. A simple model (without genetic revolution) can apparently explain the punctuated character of the palaeontological record, but cannot explain a number of other phenomena. For example, the very evolution of adaptive traits in sexually reproducing species in which, as we have shown in the previous chapters, such an evolution caused by selection should not occur. Or the "minor detail" that becomes increasingly clear in studying the occurrence of gradualistic and punctuated evolution in various types of organisms. As I mentioned above, amongst marine organisms, gradualistic evolution tends to be encountered rather more frequently in unicellular organisms and punctuated evolution rather more frequently in multicellular organisms. Simultaneously, peripatric speciation should occur in both unicellular and multicellular organisms. However, asexual reproduction is encountered far more frequently in unicellular organisms. It is thus quite possible that the more frequent occurrence of gradualistic evolution in unicellular organisms is connected with their frequent ability to reproduce asexually and thus their ability to undergo classical Darwinian evolution. Thus, unless some other explanation is put forth for the evolution of adaptive traits in sexually reproducing species and unless an explanation is found for the difference between gradualistic evolution of unicellular organisms and punctuated evolution of multicellular organisms, I would consider the rejection of the role of genetic revolution (or other population genetics phenomenon, e.g. elimination of genetic variability from a population) in explaining the punctuated character of evolution to be, at the very least, somewhat premature.

Summary and incitement

Palaeontological data, which became the stimulus for formulation of the theory of punctuated equilibria, specifically the punctuated character of evolution manifested as the almost instantaneous formation of a fully developed species and its invariability over the remainder of its existence, is fully in accord with the conclusions of the theory of frozen plasticity. In addition, if it were also confirmed in the future that sexually reproducing species undergo punctuated evolution and asexual species undergo gradualistic evolution, it would be apparent that, of the two proposed mechanisms of punctuated evolution, the original one, assuming the participation of genetic revolution, is more correct, while that which was later preferred is erroneous. In the following chapter, we will see what the results of laboratory experiments tell us about the existence of frozen plasticity.

Footnotes

1. Information on the number of dinosaur fossils was taken from the *Proceedings of the National Academy of Science, U.S.A.* 91: 6758–6763, 1994.
2. Of publications on this subject, I can recommend, e.g., *Social Studies of Science* 23: 342–362, 1993, *Science Communication* 16: 304–325, 1995.
3. Probably the best known Czech immunologist, Jan Klein, published an interesting article on this subject and on some other consequences of the "hegemony of the average" in contemporary science in *Lymphology* 18: 222–131, 1985.
4. Here a full citation is in order: Eldredge, N. & Gould, S. J. (1972), *Punctuated Equilibria: An Alternative to Phyletic Gradualism*, Schopf, T.J.M. (ed.), *Models in Paleontology*, p. 82–83. San Francisco. The first (less explicite and less provocative) article on the subject (written by N. Eldredge alone) is *Evolution* 25: 156–167, 1971.
5. Of newer articles on this subject, I can recommend, e.g., *Trends in Ecology & Evolution* 14: 72–77, 1999, *Trends in Ecology & Evolution* 16: 405–411, 2001.
6. The role of genetic revolution in peripatric speciation is described by E. Mayr in the book *Animal Species and Evolution*. Harvard University Press (Cambridge), 1963. The genetic mechanisms of evolutionary inertia of large populations are desribed by I.M. Lerner in *The Genetic Basis of Selection*. Willey (New York), 1958.
7. The theory of shifting balances (*Annual Review of Genetics* 16: 1–19, 1982) is newly described and discussed, e.g. in *Genetical Research* 61: 57–74, 1993, *Evolution* 52: 1834–1839, 1998, *Evolution* 54: 317–324, 2000, a critical view of this theory can be found, e.g. in *Evolution* 51: 643–671, 1997, *Evolution* 54: 306–317, 2000.
8. The interesting and inspiring book of S.J. Gould *The structure of evolutionary theory*. The Belknap Press of Harvard University Press (Cambridge), 2002 will probably be read only by really serious students. It is rather extensive...

CHAPTER 13 And what does genetic data have to say?

Gould and Eldredge were not geneticists and thus they followed events in this field from some distance. Simultaneously, the uninitiated observer could very easily gain the impression that the debates of theoreticians and evolutionary biologists in the middle of the last century about the importance of genetic interactions and the ability of the population to respond to selection pressures and about the role of the genetic revolution in the development of new species tend to lose sense in the light of the results obtained in modern genetics.

How should heredity be measured?

Geneticists normally perform measurements in which they measure the heritability of the individual traits and determine the percentage by which additive heritability contributes to this (see the box), or determine experimentally whether evolution occurs faster in small or large populations, or in populations with a large or small number of founders. Under these conditions, it seems superfluous to debate whether the heritability of traits exists, or not, and whether the traits of species can change in response to selection pressure without previously undergoing genetic revolution. So what do the results of genetic experiments to date tell us?

Most of the published results clearly indicate that heritability of individual traits is usually of the order of tens of percent, that evolutionary responses to external selection pressure occurs faster in large than in small populations, and that a population established by a small number of individuals does not respond better to selection pressure than a population established by a large number of founders. Although similar results are often published in the most serious professional journals and are cited even in basic textbooks in the field, it will certainly not hurt to more carefully consider the degree to which we can believe them and what they actually mean.

Box 13.1 The additive component of heritability

The heritability of traits is determined by the fraction of genetically determined variability in the given trait in the total variability of this trait, i.e. in the variability determined both genetically and nongenetically (by the effects of the external environment). Heritability in the narrow sense of the word expresses the fraction of the additive component of genetically determined variability in the total variability of the given trait. Additive variability is the component of the variability that is additive in its effects. If allele A of one gene acts, on an average, in its carriers to increase their body weight by 10% and allele B of another gene acts, on an average, in its carriers to increase their body weight by 5%, and if this is an additive component of the variability in both cases, then the carriers of alleles A and B will be, on an average, 15% larger than the carriers of other alleles. If this increase is smaller or greater than 15%, then nonadditive variability is involved.

It will be instructive to begin with the measurement of heritability. Heritability in the narrow sense of the word, i.e. the component of heritability that determines the ability of the population to respond to selection pressure and that is of decisive importance from the standpoint of evolution of the population and the species, is measured by two quite different methods. One measures the heritability from the response of the population to selection pressure and the other is based on measurement of the degree of similarity between the studied trait in parents and their offspring (in practice, mostly amongst siblings that have only one parent in common). Thus, both methods should measure the same quantity and both should provide the same result for the same trait. But they do not.[1] The results obtained by the two methods differ so fundamentally that the differences cannot be explained simply in terms of a random error in one measurement or the other. If we think more carefully about the two methods, we realize that we could not really expect any other result.

Think back to how the population responds in evolution to a constant selection pressure. Initially, the response of the population is rapid, later it becomes slower and finally it completely stops responding to the selection pressure. Of course, amongst other things, this means that the method of measuring heritability on the basis of the degree of response of the population to a selection pressure must yield a completely different result in each time interval of the experiment, in dependence on the momentary degree of deviation of the genetic composition of the population from the equilibrium (evolutionarily stable) value. From an evolutionary point of view, the heritability of the trait and thus the ability of the population to respond to selection pressure in the area around the equilibrium state is not, in general, important. If, after a small deviation from this equilibrium state, the population rapidly stops responding to the selection pressure, i.e. the heritability decreases to almost zero, the population cannot undergo classical Darwinian evolution whatever the level of heritability of the particular trait at the beginning of the experiment.

The second method also contains serious drawbacks. For example, in this type of study, the agreement between the average weight of the two parents and the average weight of their offspring is compared. Here the difficulty lies in the fact that progeny do not share with their parents only genes that affect the weight of the individual directly, but also genes that affect the weight only indirectly through affecting the influence of other genes. The genetic background, on which the individual genes act in related individuals, is similar and thus the measured heritability values must be overestimated. Geneticists are aware of this problem and try to resolve the entire problem by not comparing traits in parents and their children, but in half-siblings, i.e. in individuals that have only one parent in common. However, even this method does not resolve the whole problem. Even half-siblings share, in addition to the genes directly affecting the weight of the individual, also an above-average percentage of genes that indirectly affect the weight, in that they form the genetic background for the other genes. In short, if we were to estimate the weight of an individual on the basis of his four grandparents or even on the basis of his eight great-grandparents, the precision of our results would certainly be much lower than for the calculation based only on his two parents. Four grandparents and two parents have the same number of common alleles as the studied individual; however, together, the four grandparents have far more alleles that the studied individual does not have – the genetic background of common alleles (and thus also their manifestations) must necessarily be different in the grandparents. Nevertheless, in determining heritability, geneticists act as if they believe that the same results would be obtained in all three cases (great-grandparents, grandparents and parents) although I would not want to suggest that they really believe this.[2]

Larger is better (and so what?)

Experiments in which the rate of evolution in small and large populations is compared frequently indicate that evolution of adaptive traits occurs more readily in large populations.[3] These results could present a certain problem for the proponents of Wright's theory of shifting balances, but they do not tell us anything at all from the standpoint of the theory of frozen plasticity. The theory of frozen plasticity assumes that the evolution of new traits occurs in a period when the number of its population has already grown to the usual values, but their genetic variability has not yet reappeared.

By the way, even the proponents of the theory of shifting balances need certainly not give up the fight on the basis of the results of these experiments. As my poacher ancestors used to say: "Why take your trousers off when the buckshot is still far away?".[4] To begin with, a number of experiments have yielded quite the opposite results, i.e. showed that division of the population into several small populations actually does accelerate evolution. More importantly, in evolution, in contrast to statistics, the performance of average individuals (populations) is not decisive, but rather the performance of a few successful individuals (populations). It is quite possible that, on an average, the evolution of adaptive traits progresses faster in large, unstructured populations than in populations divided into small populations. However, if none of the large populations overcomes the very deep valley in the adaptive landscape (i.e. does not overcome the developmental stage when organisms are poorly adapted to the life style of the parent species and simultaneously are not yet completely adapted to the life style of the new species), while at least some of the separated populations can get across them to the opposite slope, development in small populations will play an overall greater role in evolution than development in large unstructured populations. (And Sewall Wright can have the last laugh and thus, as is well known, the best laugh.).[5]

Microevolution is not macroevolution

The results of experiments studying, on populations with the same size, the effect of the number of founders and thus the genetic variability on the progress and rate of evolution of adaptive traits are ambiguous. Some of the results of laboratory studies and experiments in the field indicate that evolution occurs more readily in populations with low genetic variability; a great many studies have nonetheless demonstrated the opposite. However, I am of the opinion that this type of

Box 13.2 Microevolution and macroevolution

Evolutionary processes that tend to occur at a population level are considered to be microevolutionary, while evolutionary processes (the formation and disappearance of the large branches of the phylogenetic tree, the formation of new body plans) that occur at the level of species can be considered to constitute macroevolution. Macroevolutionary processes are slow and prolonged and it is hardly possible for us to study them in our experiments. The element of chance plays a much greater role in them than in microevolutionary processes, see Chapter 4. The main source of evolutionary novelties in macroevolution consists in mutations arising in a local population, while the main source of novelties in microevolution is gene flow – the arrival of new alleles in the population through migrants.

experiment cannot basically tell us anything about the validity of the theory of frozen plasticity. We must come to terms with the fact that our experiments can study only microevolutionary, and not macroevolutionary, processes.

It is sometimes rather difficult to determine the differences between macroevolutionary processes and microevolutionary processes; however, one difference is very easy to define. Newly formed mutations are the basic factor driving macroevolutionary processes. In microevolutionary processes, which occur on a much smaller time scale and in much smaller populations, there are so few new mutations that the main material that is available for natural selection consists in the alleles that were formerly present in the local population or alleles that entered the population through migrants. Thus, if we study the microevolution of populations in our experiments, the ability of a population to respond to selection pressure is determined not only by the evolutionary plasticity of the particular population which, in accordance with the conclusions of the theory of frozen plasticity, should be larger in a genetically homogeneous population, but also, primarily, by the amount of genetic variability present in the given population at the beginning of the experiment. It is obvious that far more alleles, from which selection can choose, are present from the very beginning in genetically more diverse populations established by a greater number of founders. As a consequence, at least at the beginning (before genetic homeostasis comes into effect, see p. 135), microevolutionary processes can occur faster without regard to the lower effectiveness of selection in the genetically more variable populations of sexually reproducing species. If we really wanted to test the validity of the theory of frozen plasticity by comparing the effectiveness of selection in genetically uniform and genetically diverse populations, we would have to follow the fate of individual new mutations that we would introduce simultaneously into one or the other type of population. It follows from the theory of frozen plasticity that a much larger percentage of such introduced mutated alleles are fixed or completely removed from a genetically more uniform population. In contrast, a much larger percentage of new alleles will be retained in the population in a constant frequency in a genetically more diverse population and would thus become a permanent component of the genetic variability of the population.

Summary and incitement

To summarize. The basic assumption of the theory of frozen plasticity, i.e. greater evolutionary plasticity of a genetically more uniform population than of a genetically diverse population, can be tested experimentally; however, the arrangement of most experiments to date is not optimal for this purpose. In our experiments, we are forced to study microevolutionary processes in which the key role of the source of evolutionary novelties is played by the genetic variability present in the population at the very beginning of the experiment. In contrast, in macroevolutionary processes, which are the subject of interest of the theory of frozen plasticity, regularly arising mutations are the source of evolutionarily new features. In genetic experiments, the favourable effect of greater plasticity of a genetically uniform population can be counterbalanced by the fact that fewer alleles from which selection could choose are present (and no new alleles arise during the short-term experiment). Thus, in the future, it will be necessary to perform experiments in which the fates of individual suitable mutations introduced simultaneously into a genetically uniform and genetically diverse population are compared. In the next chapter, we will consider the results of "experiments" performed for us by nature. Specifically, we will look at evolution in species in which we would expect a higher level of evolutionary plasticity, i.e. species on marine islands, species with secondary asexuality, and self-fertilizing species.

Footnotes

1. The frequent and marked lack of agreement in the results of measurement of heritability using various methods is discussed, e.g., in the book *Genetics and Analysis of Quantitative Traits*. Lynch, M, Walsh, J.B., Sinauer Associates (Sunderland, MA), 1998 – specifically in the chapter Analysis of Short-Term Selection Experiments.
2. Sorry of the oversimplification... If you prefer to read more exact papers on related subjects, please try e.g. *Theoretical and Applied Genetics* 115: 933–944, 2007, *Genetics* 130: 195–204, 1992, *Proceedings of the Royal Society of London Series B-Biological Sciences* 260: 21–29, 1995. Interesting figures showing vanishing heritability of biological traits are available in *Journal of Evolutionary Biology* 19: 994–1002, 2006. (The authors did not comment the observed trend.)
3. The elevated ability of genetically uniform populations to respond to selection pressure was observed, e.g., in *Genetics*, 81: 163–175, 1975, *Genetics*, 114: 1191–1211, 1213–1223, 1986, *Evolution*, 43: 1800–1804, 1989, and reduced ability has been described, e.g., in *Evolution* 50: 723–733, 1996.
4. This saying should properly be "Why roll your trousers up when the ford is still far away?" However, in the Czech milieu, there is also the (erroneous) version with buckshot. The idea of a poacher who is running away from the gamekeeper in the woods and, as he is running, preventatively takes off his trousers, so they don't get full of holes when the gamekeeper shoots at him, seemed to me so marvelously absurd that I decided to support it in this way. I am curious to see whether I manage to reverse the unfavourable ratio of citations of this saying on the internet (in 2006, the ford (brod in Czech) was leading over buckshot (brok in Czech) 506 : 72).
5. Just this phenomenon, i.e. a worse ability to respond to selection on an average and simultaneously the best ability to respond to selection in some genetically uniform populations, was observed in experiments on flour beetles *Tribolium castaneum*. *Journal of Animal Breeding and Genetics–Zeitschrift fur Tierzuchtung und Zuchtungsbiologie* 118: 181–188, 2001.

Chapter 14 Evolutionary plasticity in experiments performed by Mother Nature

As pointed out in the previous chapter, it is rather difficult in short laboratory experiments to study the effect of genetic polymorphism on the evolutionary plasticity of a species. Fortunately, nature performed a number of similar experiments long before our time, so that we have nothing more to do than subsequently analyze their results.

Why are species on oceanic islands "strange"?

Oceanic islands are renowned natural laboratories. If the islands are located sufficiently far away from the mainland, each of them constitutes an independent experiment that allows us to follow the progress or at least the result of the evolutionary process. In the past, islands in the ocean were colonized by individual immigrants from the mainland or from other islands. There is a rather low probability that a species will successfully colonize an island. On the other hand, there is a quite high probability that an island species will become extinct over time. Islands last sufficiently long that it is quite probable that some of their species will undergo repeated speciation – gradually splitting off of new species. As there are substantially fewer species on the individual islands than on the mainland, there are usually many unoccupied ecological niches that immigrant species or species that can develop from them can occupy. Hence, on islands we frequently encounter the phenomenon of **adaptive radiation**, where a certain species undergoes multiple speciation and its daughter species occupy diverse niches used by the members of mutually unrelated taxons on the mainland.

As the distance of the islands from the mainland automatically creates a strong **reproductive**

Box 14.1 Continental and oceanic islands

In the past, some islands (continental islands) formed part of the mainland (or of the continental shelf), from which they either became separated when the sea level rose (British Isles) or crumbled off the edges when the continental blocks rifted (some of the Seychelles). Some islands are so large and geologically old that they basically form small continents (New Guinea). Distinctive fauna and flora occur on all types of islands, frequently including species whose relatives have become extinct on the parent mainland. From the standpoint of study of evolutionary plasticity, however, oceanic islands located far away from the mainland are important; these were formed, e.g., as a result of volcanic activity or a combination of volcanoes and corral reefs, and were colonized in the past only by individual "shipwrecked" species arriving from the distant mainland (Hawaiian Islands). Only these species underwent a dramatic decrease in the population size that could renew their evolutionary plasticity.

Box 14.2 Adaptive radiation

Adaptive radiation is rapidly repeated speciation of the species of a particular evolutionary line. It has at least two different reasons. The first cause is the penetration of the representatives of a particular line into an environment that has many utilizable but, at the given time, unutilized resources. A species that enters such an environment in that, for example, it reaches an island, "learns" to utilize the individual resources and diversifies into a great many species through gradual adaptation to the individual types of resources and individual types of environment. The second cause of evolutionary radiation is the formation of fundamentally new features that, for some reason, open a broader range of so-far unused niches for their carriers. For example, the formation of wings and the ability to fly enables the particular group of vertebrates to utilize various types of resources with scattered occurrence over a large territory. As a consequence, thousands of species of birds could evolve relatively rapidly, using various types of resources as food, from seeds and fruit through insects, to vertebrates.

isolation barrier, it is clear that speciation occurs very readily on islands and that a large number of species are formed. On the basis of the theory of frozen evolution, we would expect that the high genetic uniformity of the island populations, frequently based on a few immigrants or even on a single maternal individual, would provide island species with greater evolutionary plasticity than mainland species.[1] The most derived forms of organisms (with the greatest number of new evolutionary features) should be found here. The appearance and behaviour of species occurring on islands should differ substantially more from the typical representatives of the relevant phylogenetic lines than species occurring on the mainland. New and very different species should be formed on islands much faster than on the mainland.

The available data apparently confirm these presumptions quite well.[2] Ocean islands occupy a relatively small area of dry land and usually have relatively few types of environment and thus numbers of niches. Nonetheless, it seems that highly derived representatives of individual taxons, frequently with quite bizarre appearances, physiologies and living habits, are encountered on islands. Dormice the size of rabbits, miniature elephants or hippopotamuses are found there (of course, very often only their bones have been preserved to the present day).

Strange island forms are also encountered amongst reptiles, including giant turtles and iguanas, living off sea weed. We can even find species in groups of insects that, compared with mainland species, have much stranger shapes and frequently unusual living habits. Offhand, numerous species of fruit flies can be named, with quite unbelievable shapes, from Hawaii[3], or gigantic stick insects from Lord Howe Island. Deep ocean trenches are similar to islands, as their inhabitants are adapted to the high pressure and thus have only minimal possibilities of moving from one trench to another. The fauna in these marine "islands", for example, deep-ocean fish, are again quite bizarre. This would not be so strange as this could be a result of adaptation to the extreme conditions of their environment, high pressure and darkness. However, what cannot be explained so simply is the fact that the deep-ocean fish in the individual areas differ drastically from one another and each of them is bizarre in quite a different way.

It is, of course, quite possible that the occurrence of the appearance and living habits of

Box 14.3 Island gigantism and nanism (dwarfism)

On islands, it occurs relatively frequently that large species of animals tend to get smaller and small species tend to get larger. The usual evolutionary explanation is that, on the mainland, the greater interspecies competition or pressure of carnivores forces them beyond the limits of their optimum size, i.e. all their members tend to be large, so that they can better resist carnivores, or small, so that they can hide better and manage with a small amount of available resources. When they reach an island, where their natural enemies or competitors are absent, they can return to their optimum size – i.e. get larger or smaller to the size at which their body functions best. This could be the right explanation. However, it is necessary to consider that not every change in body structure that we encounter on islands has the nature of gigantism or nanism and, in addition, a great many long-term isolated populations in areas with low intensity of inter-species competition are apparently also formed on the mainland (however, mostly by splitting off of part of a large population), without gigantic or dwarf forms occurring to the same degree as on islands.

untypical forms or organisms is caused at least partly by the fact that the number of islands is much greater than the number of continents, so that a greater number of evolutionary "experiments" could take place here. On the other hand, islands are much younger and evolution had far less time for its experiments here. Serious comparative studies in which a comparison was made of the number of new evolutionary features in species forming in ocean islands and on the mainland or on islands formed by breaking off from the mainland have not been carried out yet.[4] However, simply leafing through an atlas of an arbitrary group of organisms shows that the strangest members of most groups of organisms currently occur or recently occurred (until the appearance of man[5]) on islands.

Asexual species – slower but better

Further data that will allow us to assess the effect of the genetic variability of the founding population on evolutionary plasticity is provided by study of secondarily asexual species. Especially amongst a great many groups of plants or invertebrate animals and, less frequently, amongst some groups of fish, amphibians and reptiles, we encounter species that, in contrast to their close relatives, stopped reproducing sexually and began to reproduce asexually. In some cases, the sex cells of females develop without any fertilization; in other cases, embryos develop only after fertilization of female sex cells by male sex cells, but the genetic material from male sex cells is subsequently rejected. However, a number of various alternatives are also encountered. From the standpoint of a physiologist, some of these means of reproduction tend to have the character of sexual reproduction, as the individual is formed from sex cells. However, from the standpoint of an evolutionary biologist, asexual reproduction is involved in all these cases.

As we mentioned in the chapter concerned with speciation, the formation of asexually reproducing lines, most frequently associated with the formation of polyploids (multiplication of the number of chromosome sets), represents only one of the frequent mechanisms of instantaneous speciation. Polyploid individuals can usually cross together but can frequently not cross with members of the original diploid species. They frequently change to asexual reproduction.[6] The original evolutionary plasticity should be renewed in asexually reproducing species (the progeny inherit the genotype of the mother) and the species should be capable of responding better to selection pressure than the parent species. It is actually known that secondarily asexual species are usually better able to adapt to extreme types of environment than the parent species. They occur much more frequently at places with, for example, high concentrations of heavy metals in the soil, at higher altitudes or in areas with extreme meteorological conditions in the cold weather zones and at higher latitudes.

As old evolutionary lines are practically not found amongst secondarily asexual organisms, at the most, groups of closely related species or, exceptionally, genera (bdelloid rotifers, an ancient group of about 300 mostly asexual species, are the proverbial exception that, as is well known, confirms the rule) it is apparent that, from the long term point of view, this type of reproduction is probably disadvantageous, even if it provides the species with greater evolutionary plasticity.

It is quite possible that every asexually reproducing species finally pays for its opportunist evolutionary strategy. If an asexual species gradually adapts to any, even temporary, change in natural conditions, it will sooner or later end up in a situation where it adapts perfectly to a temporary drastic change in conditions. If, subsequently, the conditions return to normal too rapidly (e.g. during a single generation), it has no chance of rapidly readjusting back to the original conditions. Its population must wait for new mutations and will most probably die out during this waiting period. In contrast, sexually reproducing species can never adapt perfectly to altered conditions and always retain at least part of the alleles that were suitable under other conditions so that, when conditions return to normal, they have sufficient original genetic material required for microevolutionary adaptation. Thus, if we study species in a single time plane, for example the

Box 14.4 Exceptions from rules

Rules (laws) are generally not one hundred percent valid in science and there are frequently a number of exceptions. This is caused primarily by the fact that, as they are formulated, they are excessively simplified and thus imprecise. Take, for example, the rule that females generally select males is valid only assuming that females invest more valuable resources into reproduction than males. However, this is not true for a number of species. For example, it does not hold for giant water bugs of the Belostomatidae subfamily, where the female lays her eggs on the back of the male and he then carries them, defends them and ensures that they get enough oxygen for three weeks. The total weight of the eggs is twice that of the male and care for them is a great burden on him. Females can copulate with a number of males, but a male decides whether he will accept eggs from a female or not. As the area of the backs of males is a factor limiting reproduction, fierce competition occurs amongst females for males willing to accept batches of eggs.[7] The adage "the exception proves the rule" should properly be "the exception tests (or allows testing of) the rule". In this form, it is a profound truth – when we study the individual exceptions from rules, we should always discover their cause (in the above case, the cause of the deviation from the rule is the fact that the males, and not females, invest the more valuable resources into offspring). If we can find the reason for the individual exceptions, we confirm that we have understood the nature of the rule properly.

present, then the evolutionarily plastic asexual species must necessarily seem more evolutionarily successful (i.e. more capable of adapting to more diverse conditions). However, if we follow the fate of species over a larger time scale, for example on a macroevolutionary scale of millions of years, the evolutionarily freezing or, more exactly, the evolutionarily "elastic" sexually reproducing species are found to be more evolutionarily successful.

Why wheat is "spoiled" faster than rye – microevolution of self-pollinating species

Similar differences in evolutionary plasticity to those found between asexual and sexual species could also be expected to a smaller degree between self-pollinating and cross-pollinating species. If the macrogametes (eggs) of a certain individual are fertilized only by microgametes (in plants by pollen, in animals, sperm) of the same individual, for example in self-pollinating plants, then we would be more inclined to expect evolutionary plasticity of the species than if the macrogametes and microgametes come from two different members of the same species, as in cross-pollinating plants. This fact can have a considerable effect on plant breeding praxis practices. If we compare old sales catalogues of companies selling seeds at time intervals of, say, decades up to the present time, it is said (I have not done this, so I cannot guarantee that this is true) that the individual varieties of wheat change very rapidly here. In contrast, rye is quite stable and the same varieties are apparently sold and grown over long periods of time. Simultaneously, wheat differs from rye in that it is self-pollinating. If a farmer attempts to maintain a certain self-pollinating species over a long period of time, it is highly probable that its useful value will decrease from one season to the next. Natural selection acts constantly on the plants, which improves the viability and fertility of the plants at the expense of their useful value. The properties of varieties that are useful for man need not be useful at all for the plant itself. This is also true for cross-pollinating varieties or cross-pollinating species and here natural selection also "attempts" to increase the viability and fertility of the plants at the expense of their usefulness. However, because of the lower evolutionary plasticity of cross-pollinating species, the variety does not respond to this selective pressure and its useful properties do not deteriorate over time. Of course, at the present time, practically all seed material is prepared by crossing two different maternal lines, so that these phenomena have apparently ceased to apply to seed material. But who knows – both maternal lines must be maintained over long times by the company supplying the seed material and this can constitute a problem for self-pollinating species.

In the light of this, it would perhaps be interesting to recall a practice, at first sight very suspicious, that was recommended in the 1930s by Soviet Lysenkoists. (Here, I am not thinking about the practice of sending scientific competitors to labour camps for re-education or starvation. From a theoretical standpoint, it would be hard to think up anything against these techniques. They certainly fulfilled their purpose very well in Stalinist Russia and would certainly fulfil their purpose very well at the present time. However, in our less drastic times, they have gradually been replaced by less drastic and less effective, however, in principle, well-functioning methods based

Box 14.5 Trofim Denisovich Lysenko (1898–1976)

A capable or rather capable-of-anything demagogue and absolutely incapable plant physiologist who, in the 1930s and 1940s, got on Stalin's good side and, shielded by his absolute political power, for a great many years and in the name of Marxism-Leninism, practically destroyed first genetics and later other fields of biology in the Soviet Union and partly in its political satellites. The era of Lysenkoism in the Soviet Union finally ended in the 1960s. He declared that genetics at that time was a bourgeois quasi-science serving the interests of the governing capitalist class and initiated its replacement by progressive, Soviet genetics. Lysenkoists stated that there are no genes, that the hereditary properties of organisms change under the influence of natural conditions, that one known species can change into a different known species, e.g. as a consequence of lack of nutrition, or that living cells can be formed in a test tube from a mixture of simple substances. In the Soviet Union and other communist countries, they found a great many willing helpers who, partly from fear, partly from stupidity and partly from calculation, massively falsified scientific data and liquidated any scientific opponents through political means. In their campaign against official genetics, they brought a number of interesting phenomena to light that were known by older breeders and that seemed to be contrary to accepted genetic knowledge of the time. In this way, they discredited them for a long time and prevented them from becoming the subject of serious study.

on minor daily intrigues.) The suspicious practice I have in mind consists in the strange agrotechnical methods that were strongly recommended in the works of Trofim Denisovich Lysenko.

This consisted in the technique of cross-pollination of wheat in the framework of a single variety that would prevent gradual deterioration of the quality of the seed material. At first glance, this recommendation seems completely ridiculous and it is very probable that the economic benefit of the technique, consisting of tearing out the stamens from the individual heads of wheat using special tweezers, was negative. However, it is quite possible that preparation of seed material using enforced cross-pollination could really stop the deterioration in the useful properties of self-pollinating varieties, as it would substantially reduce their evolutionary plasticity and thus their ability to respond to natural selection. In fact, it would probably be possible to breed cross-pollinating varieties whose useful properties would not gradually deteriorate as a result of evolutionary freezing. And this could certainly be economically more interesting.

The wise naivety of Sir John Sebright

While we are on the subject of breeding: in his book on the variability of plants and animals[8], Darwin expressed his surprise at a very strange technique used by the breeders and keepers of the time. In order to prevent deterioration of their breeds (i.e. to prevent natural selection from improving the average viability or fertility of animals in the herd at the expense of their useful properties), better-off breeders and improvers kept two herds of individual breeds, if possible, under as different conditions as possible, for example one in the mountains and one in the lowlands. From time to time, they crossed animals from the two herds. The **Sir John Sebright** effect appeared in the new-born crosses and their progeny, i.e. renewal of the original useful properties of the breed.[9]

The breeders of the Darwinian and Pre-Darwinian periods can be forgiven for using these nonscientific techniques. Mendel's laws were rediscovered only at the beginning of the 20th century, so our breeders could not have known that their method is erroneous from the scientific point of view and cannot (must not) work. The problem lies in the fact that breeders and improvers continued to use and probably still use this method long after it became apparent that the natural conditions under which the herds are kept cannot in any way affect the genetic material of the animals and that crossing of animals from herds kept under different conditions has no sense at all from the viewpoint of modern genetics. It is quite possible that breeders continue to employ methods introduced long ago simply from inertia. However, it is also quite possible that nature ignores the knowledge of modern genetics and that the method actually works. From the viewpoint of the theory of frozen plasticity, the given breeding procedure is certainly not as ridiculous as it seems from the standpoint of classical genetics. In two different environments, microevolutionary processes occur through the action of different selection pressures in different directions. Thus, in each herd, the frequency of different alleles increases or decreases through the action of natural selection. If, once in a while, we mix the gene pools of the two herds, the original frequency is renewed to a certain degree and thus the original properties, useful for human beings, are also renewed.

Selection in us and inheritance of acquired traits

In some types of organisms, e.g. plants, the Sir John Sebright effect can also appear at the level of the individual. Some plants are purely cross-pollinating. If they are pollinated by pollen from the same plant or pollen from a plant obtained by vegetative cloning (e.g. grown from the other half of a divided rhizome), they do not produce any seeds. In older genetics textbooks, we can encounter the statement that the pollen incompatibility of cloned plants can be overcome by growing each of the clones under as different conditions as possible, for example, one under wet conditions and one under dry conditions.[10] I should add that these experiments were mostly performed during the era of Lysenkoism in the Soviet Union. Thus, it is not clear how far we can believe them. I would think that someone who arranges that those with different opinions will starve will similarly be willing to "cook" the results of his experiments. On the other hand, it is not so difficult to carry out this experiment. I would say that there is a 50:50 probability that some cloned plants (beets are mentioned in this connection in the literature) will actually behave as described in older textbooks. As the plants grow, apparently individual cell lines are selected. While the cells of these lines have the same genotype, as they originally developed from a single germ cell, their chromosomes differ as a consequence of mitotic recombination and epigenetic modification of DNA and chromatin. It is not necessary to consider the meaning of mitotic recombination, epigenetic modification and chromatin too deeply. It is sufficient if you believe me that the way in which the individual genes affect the phenotype of their carrier depends on their order and their surroundings (nearby genes) on the chromosomes and on chemical groups that are attached to them or to the proteins that encase them.[11] Two cloned plants grown under various conditions will have the same genotype; however, as a consequence of selection of different cell lines in their tissues, the phenotypes will differ far more than those of two cloned plants grown under identical conditions. Of course, it depends on how pollen compatibility is controlled in the given species; however, in some systems, genetic or epigenetic differences of two plants with identical genotypes can result in partial or even complete pollen compatibility and thus enable their reproduction.

Why are identical twins identical?

This is in no way related to the theory of frozen plasticity, but the reader might wonder how clonal organisms, for example identical twins, can be so similar when, during their lives, similar to plants grown from a cut-up beet, they independently accumulate various mitotic recombinations and epigenetic modifications.

Basically, two answers can be given to this question. Identical twins are frequently not as identical as we would expect on the basis of the identity of their genotypes. Under normal conditions, identical twins, as least in humans, share not only genes, but also similar or even identical environments. When twins are separated soon after birth and brought up separately in very different environments, it is sometimes very difficult to believe when they are adults that they are identical twins (or that they are even siblings).

The second reason why the genetic diversity of somatic lines of cells has a relatively small effect on the appearance and properties of organisms under normal circumstances is the fact that the properties of the individual cell lines "average out" together in the framework of the tissues and the entire organism. The individual cells in the tissue can differently "turn on and off" their genes as a consequence of recombination that occurs (with low frequency) during mitosis or epigenetic changes (chemical modifications of the various areas of the DNA or chromosomal proteins); however, in the framework of the tissue, this is not greatly manifested, as random deviations from the original properties in individual cells cancel one another out. In plants, this need not even be manifested in the properties of the cells themselves, as they exchange macromolecules, RNA and proteins amongst one another, as well as low- and medium-molecular weight substances, the products of their metabolism.[12] Thus, the properties of individual cells are determined not so much by their actual genetic and epigenetic apparatus, but by the genetic and epigenetic apparatus of all the cells in the particular tissue. It is, in fact, possible that this is an evolutionary adaptation intended to prevent competition between the individual cells and the individual cell lines in the body of the organism. If the properties of the individual cells, including the rate of growth, are determined by the properties of the entire cell population and not by their own genetic and epigenetic apparatus, there is a substantially lower danger in a multicellular organism that selfish clones would, in time, predominate, that they would prefer their own reproduction over the interests of the multicellular organism and that the individual would thus suffer from cancer.

As is almost always the rule in biology, this mechanism also does not work one hundred percent and the genetic information contained in the nucleus of the cell affects certain properties of the cell so that certain selection of cellular clones occurs anyway under these conditions. In plants, i.e. organisms in which the heredity of acquired properties is most frequently described, competition between the individual parts of their bodies, most frequently apparent between individual branches, also contributes to differentiation in the individual through the mechanism of selection amongst the individual cells of a single organism. Branches well adapted to the local conditions prosper, while the poorly adapted wither away and die. If there were no exchange of molecules between cells and thus averaging of the properties of the cells in the tissue, the intensity of selection within the organism would be even greater.

Grafting of tomato plants and the great fraud that maybe wasn't a fraud after all

At the end of the chapter, a little side turn from the side turn. It does not belong here, but some authors, like me, are terribly undisciplined. The averaging of the properties of cells in the framework of plant tissues can explain another class of results mentioned in the publications both of Lysenkoists and of Japanese geneticists. This consists in the transfer of hereditary traits from the stock to the graft. It was found in experiments that, following grafting of yellow tomatoes on a stock plant with red tomatoes, the graft also started producing reddish tomatoes. This would not be surprising; it can be easily imagined that metabolites from the stock diffuse or are even actively transferred to the graft and can affect the appearance of the fruit.

However, it is very surprising that the seeds of the reddish fruit produced plants that were also reddish. Some authors stated that repetition of this experiment, i.e. grafting the newly grown plants on the red-fruited variety, eventually leads to the production of the occasional red fruit amongst the reddish tomatoes. And the plants grown from the seeds of these red fruit were apparently also red-fruiting. Basically, Japanese geneticists obtained the same results in their experiments, performed on eggplants.[13]

Box 14.6 Metabolites

Metabolites are the products and intermediates of metabolism. Metabolism has two components, catabolism and anabolism. In catabolic processes, substances originally derived from food are converted in a great many processes, which are catalyzed by the individual enzymes, into simpler substances – the building blocks of the bodies of organisms and waste products leaving the organism and entering the environment. In the process of catabolism, a useable form of energy is also produced, which is subsequently consumed in all the life processes. In the processes of anabolism, more complex molecules, from which the body of the organism is formed, are created from the simple building blocks. In autotrophic organisms, such as green plants, the simplest inorganic substances form the building blocks for the synthesis of more complex molecules; these include carbon dioxide and water and the necessary energy is derived from solar radiation. In heterotrophic organisms, such as animals, the building blocks and the energy required for catabolic processes are derived from the organic substances contained in foods.

At first glance, it would seem that there are only two possible explanations for this phenomenon. The first explanation is ridiculously simple – that this is a scientific fraud and the relevant data and perhaps the entire experiments were thought up by their authors. Although scientists falsify their results far less than they could, cases of falsified data are also encountered at the present time. In my, perhaps rather cynical, opinion, the relative rarity of falsified data is not caused so much by the high moral quality of scientific workers, but far more by the fact that it does not pay. If someone falsifies his results in that he "manufactures" an important discovery, for example, obtaining the clones of stem cells, it will be very soon discovered that his results are not completely in order. The falsifier can enjoy his five minutes of fame, but that is probably the end of his career. By the way, he will usually not even be able to enjoy his five minutes of fame – it is very difficult to publish any significant results and, consequently, more experienced authors frequently attempt to mask the real significance of their discoveries from malicious or excessively careful reviewers. In any case, this follows from the results of studies monitoring the publication fates of a number of important discoveries in the previous century. The more fundamental a discovery, the more difficult it was to get it published. Amongst discoveries that later even resulted in the awarding of the Nobel Prize, it was no exception for them to be rejected by as many as ten journals before being accepted for publication.[14]

If a falsifier tries to "manufacture" trivial data, he is not much better off. Trivial data, for example expected results, are generally much easier to publish than a fundamental discovery. On the other hand, it is not worthwhile falsifying them. As a rule, it is generally much easier to measure trivial results than to think them up so that no one will realize it. In short, trivial results are lying about everywhere, it is sufficient to pick them up and it is not necessary to risk one's scientific career for them. Understandably, in the Soviet Union of Stalin's era, things were rather different. Under conditions where the main precondition for a successful career was a willingness to declare that black is actually white and that one plus one equals an arbitrary number depending on the latest decision of the party bodies, the motivation to think up the strangest data (understandably, only those that were in accordance with the official opinions of the bosses of the time) was very high.

Conditions are understandably better in Japanese laboratories, where I was able to spend somewhat more than one year. However, I have the feeling that, even here, the risk of falsification of data is relatively high. The personal motivation of the workers in laboratories to obtain just the results that their "sensey", i.e. beloved teacher, supervisor and the spiritual father of the team, (apparently) wants is so high that the diligent research worker or laboratory technician sees precisely these results. In any case, if it is finally shown that things are quite different, there is still a final honorable solution – seppuku (for the uninitiated – this is the correct word for hara-kiri). This is, understandably, mostly a joke – research workers and laboratory technicians do not now commonly slice open their bellies; nonetheless, suicides are perceived quite differently (more positively) in Japan than, e.g., in the Christian (post-Christian?) world, so this is a means through which a great deal can still be remedied.

Box 14.7 Viruses and retroviruses

Viruses are representatives of a noncellular form of life. They consist basically of genes or, rather, mutually cooperating groups of genes that have freed themselves from their original genomes and adopted a parasitic way of life. They employ the molecular apparatus of the host cells to reproduce their nucleic acids. Some viruses are formed basically of a chain of naked nucleic acid, while others have a protein coating (capsid) and others are enclosed in further complicated coatings. Viruses usually employ the molecules of host cells for their life functions. However, they frequently bring their own protein molecules into the cells, or carry these molecules coded in their nucleic acid. The genome of some viruses consists of several genes while, in others, the genome is enormous and consists of several hundred genes.[15] As viruses are relatively self-reliant biological units (their reproduction is not dependent on reproduction of the host organism) capable of undergoing biological evolution, they can be considered to be living organisms. (Something that is capable of undergoing biological evolution either is an organism or will develop into an organism over time.) Retroviruses, which include the cause of AIDS, are a large group of viruses whose genome is formed by RNA (ribonucleic acid). This RNA is first transcribed in the host cell by a special enzyme (RNA-dependent DNA synthase) to DNA and inserted in the chromosome of the host cell. Retroviruses frequently carry the genes of the original host cell in their genome and can thus mediate in the transfer of genes from one species into the genome of another species. Consequently, in the past, they could have played an important role in the evolution of the genomes of other organisms.

The other explanation of the transfer of the genetically determined red colour of tomatoes from the stock to the graft takes into consideration the transfer of genetic information (part of the DNA, or rather RNA) from the cells of the stock to the nuclei of the cells of the graft. This is probably technically possible, either directly or through a virus or retrovirus.

However, I must say that this possibility does not seem very probable to me. It is not clear to me how and why the genes responsible for red fruit colour would be transferred from the stock, i.e. just the trait that the experimenters were studying – this would have to involve a massive transfer of genes and it is not clear to me how the relevant gene could be inserted into the proper position in the genome. This could probably be determined using gene conversion (see Box 14.8) and it is true that some results published in 2005 in the journal *Nature* indicate that a plant could actually have a copy of at least some of its genes hidden somewhere. An allele that was present in the predecessor of a plant could unexpectedly return to the relevant site on the chromosome after several generations.[16]

In case of transfer of red colour from the stock to the graft, however, this explanation does not seem to me to be too probable. As was mentioned above, similar experiments were described by Japanese scientists using eggplants and it would have to be a great stroke of luck for the relevant genes to be transferred in two different systems.

So, what other, if possible more probable, solutions remain? I would bet on **visualization of masked genetic variability that was already present within a single plant**. Under normal conditions, the individual fruits of a plant are similar not because their cells are genetically identical, but to a considerable degree because they are on a single plant and, because of transport of molecules within the plant, the properties of their cells are averaged. If a graft of a yellow-fruit plant is grafted onto a red-fruit plant, molecules derived from the red-fruit variety begin to enter

Box 14.8 Gene conversion and molecular drive

Gene conversion is repair of the nucleotide sequence of one allele according to the sequence of another allele occurring in the same cell. Gene conversion is a very frequent phenomenon, as it is part of the normal process of genetic recombination. However, generally, we do not know about this, as most variants of genes have the same probability of acting as a pattern for repair of another variant of the gene as of acting as the object of the repair. However, some alleles act far more frequently as a pattern for repair and, as a consequence of the process of gene conversion, can spread rapidly in the gene pool of the population (without providing their carriers with any advantage at all). Genes that are capable of reproducing more frequently (and are thus inserted into new sites on the chromosome) spread similarly. These and similar processes that very substantially affect evolution at the level of the DNA are called **molecular drive**.

the cells of the graft and thus shift the appearance of the fruit towards red. Of course, the most shifted appearance will be exhibited by the fruit that, because of the relevant somatic mutations or somatic recombinations, will be inclined to form a red colour, for example will have activated one of the enzymes of the metabolic pathway leading to synthesis of this colour.

Consequently, grafting onto a red-fruit variety and provision of the remaining necessary metabolites simply made visible the formerly existing genetic variability between the individual parts of the plant and thus permitted the experimenter to gradually select red-fruit tomatoes on a yellow-fruit variety.

Box 14.9 Somatic mutation and somatic recombination

In most species of contemporary multicellular organisms, the body contains two fundamentally different cell lines, the germinal (germ) line and the somatic line, between which there is a sharp boundary. This boundary, called the **Weismann barrier**, is formed in some types of organisms, e.g. in vertebrates and insects, in the very early stages of formation of the embryo. The germinal line subsequently leads to the formation of sex cells, i.e. gametes, so that all the changes that occur in the germinal line during the life of the individual are transferred to the progeny. In contrast, the somatic line leads to the formation of all the other tissues and genetic changes that occur in the members of this line cannot be transferred to the progeny – they disappear with the disappearance of the individual. Of course, mutations and even recombinations occur in the cells of the somatic line. For example, somatic mutations can lead to the formation of cancer cells whose progeny – a tumour – can endanger the health and life of the individual. Somatic (mitotic) recombinations are two to three orders of magnitude less common than the recombinations that occur in the cells of the germinal line during meiosis. Their biological importance is not clear. It is possible that defence against somatic mutations and recombinations (and the subsequent selection within the organism, leading to selection of a line of selfish cells, which could endanger the functioning of the multicellular organism) could be the reason for the evolutionary formation of the Weismann barrier in organisms whose cells can travel within the organism.[17] In selection within the organism, it is highly probable that clones of cells that would reproduce rapidly but would simultaneously cease to fulfill their original function would probably be successful. In groups of organisms in which the cells cannot travel, because of the existence of interconnected cell walls, and can thus not endanger the integrity of the multicellular organism (in plants and fungi), the boundary between the somatic and germinal line is not as strict or develops much later in embryogenesis. Frequently, this boundary does not exist at all and sex organs can subsequently differentiate from cells of the somatic line practically anywhere in the body of the organism. In this kind of organisms (e.g. in tree species) somatic mutations and recombinations can play an important role, e.g. in adaptation of the organism to the local conditions of the environment.

How to make (strange) flies

The same mechanism is probably responsible for the much better known phenomenon of **genetic assimilation.**[18] The important American geneticist, Conrad Waddington, devoted a large part of his scientific life to studying genetic assimilation. Mutant flies appear from time to time in populations of fruit flies, with strangely altered venation in their wings. Under normal conditions, the frequency of formation of these mutants is very low. However, if the pupae of fruit flies are exposed to a thermal shock at a certain stage in development, i.e. to a temporary increase in temperature, phenocopies appear amongst the emerging flies, i.e. individuals that have a similar appearance (phenotype) as the mutants. However, in contrast to the mutants, the individuals bearing a phenocopy will not pass the altered wing venation morphology on to their progeny. If a population of fruit flies is exposed to the relevant selection pressure, i.e. if we allow only those individuals responding to the thermal shock by the relevant change in venation to reproduce, phenocopies will appear with increasing frequency in subsequent generations, i.e. the flies will respond increasingly willingly to the thermal shock. It is not surprising that, after some time, 100 % of the population will respond to the thermal shock through the formation of phenocopies. However, what is very surprising and what caused a warning raising of the eyebrows amongst most of Waddington's colleagues, was the fact that a number of the flies in this selected population began to form the particular phenotype even when they were not exposed to a thermal shock in the pupal stage. According to Waddington, genetic assimilation occurred here; properties that were originally caused by an external intervention became genetically determined in a very few generations.

The most probable and, in my opinion, the only explanation for genetic assimilation is visualizing of the genetic variability formerly present in the population. At the very beginning of the experiment, some flies had a genetically determined tendency to form the relevant change.[19] This tendency was reflected only in the fact that they responded to the action of a thermal shock through the formation of a phenocopy. The thermal shock only made visible the variability present and permitted Waddington to recognize and thus select fruit flies with the relevant mutation. If these individuals were then crossed and gradually selected on the basis of increasing willingness to form phenocopies, mutants or recombinants appeared in the population over time that were capable of forming the relevant change in wing venation even without external intervention. This thus does not constitute a mysterious heredity of acquired traits (as Lamarckists would like to think), but normal Darwinian evolution of a new trait by the mechanism of classical artificial selection. And thus, for a little while, we can mark the end of all the side turns from side turns.

Summary and incitement

Now the traditional summary for the reader, who lost his way through the maze of turn-offs from turn-offs. Macroevolutionary processes can hardly be studied using laboratory experiments; however, they can be studied using the experiments that nature itself carried out in the past. Study of species that developed on ocean islands and species with nonsexual reproduction or at least self-pollinating species indicates that the formation of species from small genetically identical populations and asexual species is actually faster, in accordance with the conclusions of the theory of frozen plasticity, and progresses further than the evolution of species formed in other ways or evolution of sexual (or cross-pollinated) species. The next chapter will be concerned with the ecological consequences following from the theory of frozen plasticity, specifically the effect of frozen plasticity in maintaining biodiversity. In this connection, we will speak about the special ecological features of asexually reproducing species and about invasive species. And, of course, you can "look forward" to a number of subjects that are only loosely related turn-offs.

Footnotes

1. It is not entirely clear how large the populations usually are that lead to the formation of a new species (i.e. settling an island). Apparently, there is no fixed rule here. In the extreme case, it could really involve only a single fertilized female; however, there is very good genetic documentation indicating that the founding population must have been quite large. Results frequently indicate that, for example, polymorphism in MHC genes is transferred across the boundary between species, which means that the new species was established by a greater number of individuals (carrying more than two alleles together). *Nature* 335: 265–267, 1988.
2. There is apparently no doubt about the fact that island species are very bizarre. However, opinions can differ on how frequently the observed bizarre features can be included under phenomena of island gigantism and nanism and, in this respect, whether the origin of these strange features can be explained on the basis of known phenomena – lower number of competitors and natural enemies on islands. A number of my colleagues with whom I discussed this aspect are inclined to think that the limited effect of competitors and enemies is, in itself, a factor that is capable of explaining the formation of the observed strangeness in the body structure and in the ways of life of island species. On the other hand, a number of them admitted that other factors probably play a role, at least in some groups. I am of the opinion that this question cannot be resolved in a qualified manner in the absence of specially targeted studies, in which the number of derived forms of traits (apomorphy) in island and continental species are compared in the framework of individual monophyletic taxons. For the beginning, I can offer at least a few citations of works describing the bizarre features of island species: *OIKOS*: 47: 47–56, 1986, *Philosophical Transactions of the Royal Society of London Series B-Biological Sciences* 351: 785–794, 1996, *Evolution* 59: 226–233, 2005, *Journal of Biogeography* 32: 1683–1699, 2005.
3. The radiation of some groups of fauna and flora in Hawaii is described, e.g., in *Evolutionary Biology* 31: 1–53, 2000.
4. However, some comparative studies can be found. A very extensive one was performed on song birds and demonstrated that monophyletic groups of island birds are morphologically more diversified than monophyletic continental groups in most of the monitored qualities. *Nature* 438: 338–441, 2004.
5. Since the beginning of marine navigation, the fauna and flora of ocean islands have apparently been the most endangered part of the biosphere. Most known species that became extinct as a consequence of human activity lived on islands. Man killed a number of species directly; however, even more were exterminated by animals and weeds that humans introduced to the islands intentionally or unintentionally. Feral goats are probably the worst enemy of indigenous flora, while dogs and rats killed off the original fauna. When taking into consideration strange species that developed on ocean islands, we must also take into consideration species that recently became extinct there as a consequence of human activity.
6. Originally, it was quite naturally concluded that asexual reproduction was a consequence of the formation of polyploids; to be more exact, a consequence of selection pressures caused by frequent disorders in meiotic division of the cells formed by the combination of the gametes of a polyploid with the gametes of the far more common members of the maternal diploid population. Modern models tend to indicate the opposite possibility. If a line arises in the population whose members fer-

tilize their egg cells with their own microgametes or the microgametes of close relatives (which could be a stage prior to the formation of secondarily asexual reproduction), then individuals with reduced viability, infertile individuals or unviable individuals will very rapidly begin to split off (individuals with two copies of the same detrimental, i.e. lethal or semi-lethal recessive mutation). Within a very few generations, the viability and fertility of the members of this line will substantially decrease and the line will die out in competition with the maternal species. However, if this is a species with a higher level of ploidy, i.e. with a number of chromosome sets greater than two, or if the particular line manages to increase the number of chromosome sets to even three because of genome mutation, homozygotes with three lethal or three semi-lethal alleles are split off incomparably more slowly and the particular line need not die out. See *Journal of Evolutionary Biology* 17: 1084–1097, 2004.

7. You can find out all about the secret lives and loves of giant water bugs of the Belostomidae family in *Animal Behavior* 27: 716–725, 1979.
8. C. Darwin *The Variation of Animals and Plants under Domestication.* John Murray (London), 1868.
9. I named and explained the Sir John Sebright effect, described by Darwin in his book on variability within the species (see note 8), in *Rivista di Biologia-Biology Forum* 95: 259–272, 2002. In this article, a number of other phenomena described in Lysenkoist literature are "brought out into the light" and explained from the standpoint of modern genetics.
10. N.V. Turbin *Genetika a základy selekce (Genetics and Foundations of Selection)*, Přírodovědecké vydavatelství (Prague), 1952.
11. The phenotype of an organism is not determined only by the genes that it bears in its chromosomes, but also (and, I would say, primarily) by the conditions under which and the strength with which they are expressed, i.e. how intensively RNA molecules are synthesized according to its genes (and, as appropriate, proteins are synthesized according to the RNA). Expression is affected, for example, by the DNA sequence flanking the given gene, the chemical groups that are attached to the DNA forming the given gene or the proteins that are bonded to the gene or around which the DNA is wound. Epigenetic modifications (or also epigenetic mutations) are changes in the DNA-protein complex that in some way affect the phenotype of the organism (usually by affecting the expression of its genes). Some epigenetic modifications are transferred from one generation to the next, while others disappear during DNA replication or later and thus affect only the properties of a certain individual, but not of its offspring. Again, somatic recombination can change the phenotype of the cells because of the existence of a positioning effect – if a certain allele is transferred from one chromosome to the next as a result of mitotic recombination, it can start to be expressed here (transcribed to the RNA and translated to the protein), even if it was not expressed on the original chromosome. The two homological chromosomes differ not only in the variants of the individual genes, but also in the variants of their regulation elements.
12. The transfer of various molecules and integration of signals within plant tissues are discussed, for example, in *Current Biology* 9: R281–R285, 1999, *Science* 279: 1486–1487, 1998, *Trends in Plant Sciences* 4: 340–347, 1999, *Rivista di Biologia-Biology Forum* 95: 259–272, 2002.
13. Lysenko's results are described in T.D. Lysenko *Agrobiology*, pp. 279–280, 405, Brazda (Prague), 1954, and the Japanese results can be found, e.g., in *Japanese Journal of Breedings* 29: 318–323, 1979, *Japanese Journal of Breedings* 30: 83–90, 1980 and *Japanese Society Hortulan Sciences* 49: 211–216, 1980.
14. Specific cases are described in the articles in *Nature* 425: 645, 2003 and *Science Communication* 16: 304–325, 1995.
15. A description of the giant *Mimivirus*, whose 1.2 MB genome is larger than the genomes of some bacteria and codes more than 1 200 genes, can be found, e.g., in *Science* 306: 1344–1350, 2004.
16. These surprising results, indicating that a plant could have its genes "stored" somewhere, were published in *Nature* 434: 505–509, 2005. However, serious concerns about the original experimental design were published in *Nature* 443: E8, 2006. For an alternative explanation of the phenomenon see also *Plant Biology* 9: 30-31, 2007.
17. The Weismann barrier as a defence against selection within the organism is described in detail by L.W. Buss in his book *The Evolution of Individuality*. Princeton Univ. Press (Princeton, N.J.), 1987.
18. On the subject of genetic assimilation and canalization, I can recommend *Evolution* 51: 329–347, 1997, *BioEssays* 19: 257–262, 1997, *Nature* 150: 563–565, 1942 and *Nature* 183: 1654–1655, 1959.
19. Imagine that ten different genes, whose effects are more or less additive, are responsible for a particular trait (loss of a particular branch in wing venation). If, in a particular individual, the alleles causing the loss of venation are present in eight of these ten genes and the

pupa is exposed to a thermal shock at the right moment, then the venation does actually disappear and the particular (crossveinless) phenocopy is obtained. If the alleles causing the loss of venation are present in all ten genes, the venation disappears even if the pupa is not exposed to a thermal shock. At the beginning of the experiment and in fruit flies obtained from nature, the frequency of alleles causing loss of venation is rather small, so that at least eight alleles causing loss of venation occur in only a small percentage of individuals (and ten such alleles do not occur in any). However, if, for several generations, only fruit flies that reacted to a thermal shock by loss of venation are allowed to reproduce, the proportion of alleles causing loss of venation will necessarily increase in the population. After a certain time, most of the individuals in the population will have eight or more alleles determining the disappearance of venation and a large percentage of individuals will even have ten such alleles and will thus have the relevant trait even without a thermal shock.

CHAPTER 15 Ecological consequences of the theory of frozen plasticity (or farewell to the brave Darwinist world)

Why mice haven't eaten us yet

Evolutionary freezing of the vast majority of species apparently substantially contributes to maintenance of biodiversity (i.e. number of species and species variety) in nature. Selection acts with a much greater force on species with a short generation period, which generally occur in very large populations in nature. On the basis of classical Darwinist theory, we would expect that these species would easily win in the evolutionary battle with their biological opponents with longer generation times and smaller population densities. Thus, more rapidly developing species should gradually force out species that develop slowly, whether they are their food competitors or natural enemies (predators and parasites). Evolutionary plasticity of species is a great enemy of biodiversity of ecological communities. However, if all the species occurring at a given place have the same evolutionary plasticity, i.e. almost zero (as predicted by the theory of frozen plasticity), there is a much better chance that they will be able to exist in the same place in the long term. Immediately after its formation (when it is still evolutionarily plastic), each species specializes in its niche and, in the future, will not compete too much with other species by "learning" to use further resources on which other species are dependent.

Thus an enormous number of sexually reproducing species, differing not only in the breadth of their ecological niches, but also in the sizes of their populations, and rate of reproduction, can exist for long periods in individual ecological communities and thus in the entire ecosystem of the planet Earth. If the species remained evolutionarily plastic and could readily undergo Darwinian evolution, the most successful one would force out the other species using similar niches. As mentioned above, the species with the fastest rates of evolution would gradually predominate, i.e. primarily species with a fast rate of reproduction and forming large populations (the fact that these species are called r-strategists will be mentioned later). It is probably characteristic that microbes, in which sexuality plays a much smaller role than in plants and animals, exhibit these properties. Exact and especially reliable estimates of the number of species do not exist; however, it seems that the biodiversity of microbes is much lower than that of plants and animals.[1] It is also almost certain that, in the absence of tens of millions of species of sexually reproducing organisms, forming a sufficiently diverse environment for microorganisms and thus permitting the formation of an enormous number of various niches[2], this biodiversity would be much smaller.

Why bacteria haven't eaten us yet

What does the existence of asexually reproducing species mean for maintenance of biodiversity? They should remain evolutionarily plastic throughout their existence and thus should be able to undergo classical Darwinian evolution. Thus, it could be expected that they should be successful in the battle with sexually reproducing species and thus should negatively affect biological diversity, especially its component that is designated as disparity (see Box 4.1 on p. 49). In a world occupied only by asexual species, we would expect that there would be a large number of species; however, most of them would belong to a small number of (successful) developmental branches, and thus their members would probably be very similar and would almost certainly be closely related.

Let's begin by recalling that a number of classical hypotheses searching for the causes of the evolutionary success of sexually reproducing species assumed quite the opposite, i.e. the submission of asexual species in competition with sexual species as a consequence of their lower ability to respond to the evolutionary moves of their opponents (competitors and natural enemies). What is the actual situation in relation to the evolutionary plasticity of asexual and sexual species? Sexual species can obviously not have lower and simultaneously higher plasticity than their asexual competitors. And, do you know, they can! As I mentioned in Chapter 13, two factors decide on the ability to respond to selection pressures – the ability of newly formed mutants to be fixed in the population and also the amount of genetically determined variability already present in the population. **On the short-term time scales** of ecological processes, a key role is played by the amount of genetic variability already present in the population at the beginning. A population or species does not have time to wait for a suitable new mutation – time is of the greatest importance

Box 15.1 Species in sexual and asexual organisms

The most widely used definition of a species, the definition of a biological species, can obviously not be applied to asexual organisms. According to it, the largest known group of organisms that exchange amongst themselves, or at least potentially can exchange genetic information by crossing, belong in a single species. For asexual organisms we can, however, use some other definitions of a species, e.g. the definition of a typological species or an evolutionary species. Generally (although not always) a species of asexual organism is considered to correspond to the largest group of individuals that have an exclusive common ancestor (i.e. a common ancestor that is not simultaneously an ancestor of a different species) and simultaneously share an important trait that differentiates them from the members of other species. Here, the definition of a species is, to a certain degree, subjective – a taxonomist decides what is and what is not an important trait. On the other hand, species objectively exist amongst asexual organisms (at least according to the currently prevailing opinion). However, the mutual similarity of their members is maintained, not by mutual exchange of genetic information in sexual reproduction, but entirely by selection, especially combined with evolutionary draft, see p. 58. As soon as a mutation appears in a member of an asexual species that provides it with a substantial advantage over all the other individuals in the population, all the variability in all the genes occurring in the population disappears, because the descendants of this mutant will gradually prevail. This cannot happen in sexual species, only the variability in the particular gene and the adjacent genes can be eliminated, see Chapter 6.

here, i.e. which of the competitors turns up first with an effective adaptation and forces his opponent out of the game. Consequently, sexual species are capable of rapidly adapting to changes in the environment, as their members have far greater genetic variability and are capable of forming ever newer variants by crossing, i.e. ever newer combinations of alleles. In contrast, **on longer time scales**, evolutionary plasticity is determined by the ability of the population and species to fix a newly formed suitable mutation in its gene pool. Here, asexual species are evolutionarily more plastic. (We have already mentioned on p. 143 that this greater evolutionary plasticity need not always be advantageous.)

Why asexual species prefer extremes

It follows from the above, amongst other things, that sexually reproducing species should be better off in an environment rich in resources and with many competing species. This enables them to implement their advantageous ability to more rapidly respond to evolutionary pressure from their competitors. In contrast, asexually reproducing species should be better off in environments and habitats poor in resources or where the survival of most species is limited over a long period of time by unfavourable abiotic factors. Here, the rapidity of response is not important and it is more a matter of how well the species can change its phenotype in response to the requirements of the environment.

I will not hold the reader in suspense. This is exactly the situation we encounter in nature. Asexually reproducing species or asexually reproducing lines of otherwise sexually reproducing species of plants and animals are found primarily in habitats with extreme conditions – at habitats that are extremely dry, extremely cold or extremely poisonous. The proportion of asexual species increases, for example, with increasing altitude and latitude or at places where the soil contains high concentrations of poisonous heavy metals.

It could be objected that, in these cases, the unfavourable conditions reduce the size of the populations that can survive here in the long term and thus complicate the meeting of sexual partners and give an advantage to asexually reproducing species. However, the populations of many species are extremely large under these conditions and thus the search for partners should not in any way complicate their survival. In addition, some asexually reproducing species still require the combination of female and male cells for their reproduction, without which the development of the egg is not started. The females are really spiteful here – they kindly allow the males to do their work; however, either their genes are not even allowed to participate in the formation of the bodies of their progeny, or they are allowed to participate in formation of the tissues and organs of the bodies of the progeny, but are not allowed to enter the sex cells of the progeny. Pretty sneaky, huh? The female enjoys herself, in the latter case even utilizes the genetic variability of the male genes for production of variability and thus competitiveness of the progeny but, when the going gets tough, when a decision is to be made as to which genes will enter the evolutionarily immortal line of sex cells, she says "Sorry daddy, but your genes have no rights here".

A short word about ants

The females of the ant species *Wasmannia auropunctata* have it nicely worked out in this respect. The female (queen) uses the male sperm only for fertilizing eggs from which unfertile workers are hatched and produces the future queens from unfertilized (but diploid) eggs, i.e. as her genetic copy (clone). However, as it happens, things do not quite work out as the females planned. The males program some genes on their chromosome so that, in some fertilized eggs, from which workers were to be hatched, they destroy the maternal set of chromosomes. As a consequence, a haploid individual is hatched from some eggs – this is always a male in hymenopterous insects. Thus, queens and kings reproduce clonally in this kind of ant, their gene pools are quite separate and they could even be considered to be two separate species.[3]

A brief reference to mice and men

In order that I not be accused (quite justifiably) of male chauvinism, there are certainly lots of situations where males behave maliciously towards females. For example, when mouse (and apparently also human) males program (i.e. imprint) genes for some growth factors in their sperm so that they produce a large amount of the relevant growth factor in the future embryos. As a consequence of the high production of the growth factor, these embryos are larger than the embryos of males that do not carry out imprinting of their genes. Simultaneously, the growth of embryos with imprinted genes occurs at the expense of the other embryos, and even at the expense of the state of health of the mother. The male is basically acting in the sense of the saying "Après moi, le deluge". The birth of an excessively large offspring can damage the organism of the mother and thus reduce her future fertility; however, the male can have the next offspring with a different female and the present mother can have offspring with a different male. The unfortunate mother must preventatively program other genes in her eggs to protect the other embryos and herself, so that the embryos form receptors capable of capturing and destroying the growth factors produced by the fathers' genes. Thus, if a set of genes from the mother and one from the father meet in a healthy embryo, the embryo will have normal size, that is optimal from the standpoint of the entire batch of offspring. If only the genes of the father are active in the embryo, the embryo will be too large; if only the genes of the mother are active, it will be too small. These battles between female and male genes could result in some developmental disorders, for example the Angelman syndrome and the Prader-Willi syndrome in humans. And now it occurred to me – was it actually the father who got out the war axe first? What if, at the very beginning, it was the mother who, in attempting to produce the most numerous and genetically most diverse progeny (from the standpoint of spreading out the risk, it can be preferable for the female to arrange that each future progeny have a different father), programmed her genes to ensure the formation of the smallest, still viable embryos? And thus the fathers had no choice but to try to neutralize these activities of the females through its own genes. Actually, I think the opposite scenario is more probable, but who knows. As is well known, women are beasts (and men are bastards who differ only in the degree and quality of their pretences).[4]

Why asexual species prefer extremes – continued

An end to the grievances and complaints of a justly fuming human male – let us return to the greater long-term evolutionary plasticity of asexual species. Classical evolutionary theory assumed that asexual species (and, in species capable of both sexual and asexual reproduction, the asexual lines) should be less evolutionarily plastic because, compared to sexual species and lines, they have a much poorer stock of genetic variability – resources and basic structural material for evolutionary changes. The ability of asexual species to successfully survive in extreme and, simultaneously, long-term unchanging conditions, on the other hand, indicates the greater evolutionary plasticity of these species. Biologists either magnanimously overlooked this fact or explained it by stating that, compared to sexual species, asexual species are so much weaker that they are forced out by their sexually reproducing competitors from their usual biotopes into extreme biotopes.

However, this explanation is almost certainly erroneous and actually shows how misleading the use of "healthy peasant logic" and simplifying analogies can be in science. If a limited number of new settlers comes into an empty landscape, they will probably actually divide it up so that the strongest and most capable will occupy the best sites and the best resources and only less advantageous resources and sites, that are of no interest to the strong and capable, will remain for the weaker and incapable. So far, the mentioned similarity holds between the battle of apparently evolutionarily and ecologically capable sexual species and evolutionarily and ecologically weaker asexual species. However, let us continue further with our example. What happens when a large number of settlers come into the landscape or when the settlers begin to reproduce in the territory? In this case, a large number of individuals must share the better sites and their resources. The better sites then stop being better and their rich resources stop being rich and it starts to become advantageous for the members of the population of stronger and more capable individuals to penetrate into the originally less attractive places. They then begin to force the less able inhabitants, the weaker and less capable, out of these places. And things should work out similarly with asexual species in nature if the only reason why they occupy and utilize extreme biotopes were their evolutionary weakness, which does not permit them to enter into battle with sexually reproducing species in better biotopes.

However, our experience indicates that things do not work out this way and that asexual species can survive for long periods of time in extreme habitats without being forced out by related sexual species. The most probable reason for the ability of asexual species to preferentially live in extreme habitats would seem to me to be, surprise (!), their greater evolutionary plasticity, related to the Darwinian mechanism of their evolution. Because of this greater plasticity, asexual species finally adapt (slowly) even to conditions to which evolutionarily frozen sexual species are not capable of adapting.

Beware! Invasion!

The formation of **invasive species** is another ecological phenomenon that could possibly be explained from the standpoint of the theory of frozen plasticity. Invasive species are species that, in most cases, went wild for quite unknown reasons and began to spread from their originally limited area of occurrence to new areas. In some cases, a species spreads around the whole world; in other cases, its progress is stopped by a natural obstacle, such as the sea or a mountain range. In this case, man frequently inadvertently provides a helping hand by carrying it across the obstacle, either intentionally or unintentionally. "What a pretty flower – I'll try planting that on the lake at home. What is its name? Water hyacinth? What a nice name!" Invasive species frequently have the unpleasant property of substantially affecting the character of the biological community that they enter.

The invasive species of fire ant, *Solenopsis invicta*, is capable of wiping out dozens of species of local ants in newly occupied areas and of changing the size of the populations of their prey and their competitors. In the newly formed community, it frequently happens that only the invasive species and a few originally perhaps rare species can survive. Especially invasive species of plants behave in this way and are frequently capable of causing very fundamental and striking changes in communities over extensive areas (the single-species culture of the ice plant *Mesembryanthemum crystallinum* in California or cheatgrass *Bromus tectorum* in the western part of the United States are "nice" examples).

The reason why some species "went wild" and began to behave as invasive species at a certain moment is frequently completely unknown. It is relatively easy to explain cases where human beings introduced a certain species to places where it never occurred previously and where it has no natural enemies or competitors. Islands, which have relatively few species of flora and fauna, are especially susceptible to this kind of invasion. Man is responsible for the greatest number of invasions at the present time (which is quite understandable, because it is man that constantly moves about over great distances, moves not only himself but also enormous amounts of fauna and flora). The transfer of a species to a new territory (by man or otherwise) is generally a necessary condition for invasion, but is still not a sufficient condition. In the vast majority of cases, the species succumbs to competition with the local species (which are adapted to the local conditions) and dies out. Only a small fraction of introduced species are "successful". For example, European elk were introduced into New Zealand a total of 32 times and it was only the last attempt

Box 15.2 Biological communities

A biological community is a set of species that usually occur together in certain types of habitats. Terrestrial communities are usually defined by botanists, as the plant species composition decisively affects the presence or absence of other species of organisms. The character of a community is usually determined by the presence of a few key species, the disappearance of which, for example, through unsuitable intervention by man, can affect the ability of a large number of other species to survive.

that was successful and elk occupied the entire area of the southern island. Similarly, the now excessively successful starling settled in America only after the at least ninth attempts.

It cannot be overlooked that the spreading of an invasive species does not usually occur directly from the site of its original occurrence, but from a distant site which it reached secondarily at a certain moment. As has already been mentioned, this species already reached the given site several times in the past, but behaved quite differently – either it died out at the first suitable opportunity, e.g. a cold winter or dry summer, or remained for a longer time, but only in small numbers and especially in an unspreading population. Then, suddenly, it was like something happened to it and the originally inconspicuous plant or ant set out to conquer the world.[5]

A parasite is the one to blame

The most popular explanation for the emergence of an invasive species at the present time is the **hypothesis of escape from natural enemies**, primarily from the reach of parasites. It should be pointed out that, at the individual locations within its original area of occurrence, the numbers of a future invasive species (or any other species, for that matter) are basically maintained by one of two mechanisms, "from below" by a lack of resources, i.e. through a chemostatic mechanism, or "from above" – through the action of enemies, i.e. the turbostatic mechanism. These mechanisms have nothing to do with chemistry or turbines. They were named after two different types of equipment for long-term industrial cultivation of microorganisms, the chemostat and turbostat.[6] A nutrient solution flows into the cultivation vessel of the **chemostat**, in which a population of, e.g., yeast is multiplying; simultaneously, the same volume of a solution with the waste products of the yeast and part of their population flows out. If the size of the yeast population increases for some reason, it begins to take more nutrients from the environment, the nutrient concentration decreases in the vessel and the rate of multiplication of the yeast decreases because of a lack of nutrients (the rate of their dying remains constant or increases) and the size of the population decreases because of the reduced rate of reproduction. The smaller population uses up a smaller amount of nutrients, so that the concentration of nutrients increases in the vessel, the rate of multiplication increases and the size of the population again increases. Thus, an approximately constant population of yeast is maintained in the chemostat, dependent on the preset amount of nutrients in the inflowing medium. The rate of growth of organisms in the chemostat is permanently limited by the lack of certain nutrients in the nutrient solution and the quantity that determines the result of the competition for survival within the species is the **economy of reproduction**, i.e. the number of progeny that a particular individual, e.g. a particular mutant, produces per unit of nutrients consumed.

A **turbidostat** works on a different principle. The inflow of nutrient solution into the cultivation vessel is controlled, e.g., by the signal from a photocell which monitors the density of the population of organisms (turbidity) in the cultivation vessel. If the size of this population increases for any reason, the turbidity in the vessel increases, less light falls on the photocell and this sends a signal to the pump to increase the flow rate of nutrient solution into the cultivation vessel. As a consequence, two things happen simultaneously. First, the amount of nutrients flowing into the cultivation vessel increases and thus even the larger population need not go hungry and need not limit its rate of growth. Simultaneously, however, the rate of washing of organisms out of the cultivation vessel increases, so that the size of the population begins to decrease. This reduces the turbidity, the photocell sends a signal to reduce the rate of inflow of medium into the cultivation vessel and the size of the population begins to increase again. Thus, depending on the setting of the photocell, the size of the population oscillates around a certain equilibrium value. In contrast to the chemostat, organisms in a turbidostat constantly multiply at the maximum possible rate, because their reproduction is not limited at any moment by a lack of some resource. The **maximum rate of reproduction**, i.e. the number of progeny that an individual produces per unit time under ideal conditions, is the quantity in a turbidostat that decides on the result of the competition for survival within the species. Every natural population has a stable number in the long term, i.e. the size of the population in which the rate of reproduction of its members is exactly equal to the rate of dying. This is a result of the fact that each population is exposed to one of the two types of regulation. The chemostatic type of regulation is encountered wherever the growth of the organisms is limited by a **lack of a certain type of resource**. The turbidostatic mechanism is encountered where the growth of the population is **limited by the actions of parasites or predators**.

A turn-off for possible cavillers. Yes, my dear cavillers! In nature, these two types of regulation can alternate regularly or irregularly over time. Even in case of a chemostatic population, the number of individuals can be negatively affected by the activities of parasites or predators. From the standpoint of regulation and thus from the standpoint of the quantity that is the subject of natural selection (economy vs. maximum rate of reproduction), one or the other kind of feedback is important at a certain instant – a population cannot be simultaneously regulated turbidostatically and chemostatically, because this would require that the equilibrium size of the population from the turbidostatic standpoint exactly equal the equilibrium size of the population from the chemostatic standpoint. Even if this improbable situation actually occurred, the first mutant with altered economy or maximum rate of reproduction would change one or the other equilibrium size of the population. End of the turn-off for my dear cavilers.

A comment for readers with good memories who recall the old concepts of r and K strategy. My dear readers with good memories! You may recall that **r-strategists**, i.e. organisms whose populations are the first to occupy new biotopes, which produce many offspring, of which only a small percentage lives to adulthood, are organisms exposed to turbidostatic regulation and thus turbidostatic selection for the maximum rate of reproduction. In contrast, **K-strategists**, i.e. organisms whose populations predominate over r-strategists in stabilized, species-rich biotopes, producing only a few progeny during their lifetimes, of which a large percentage live to adulthood, are organisms exposed to chemostatic regulation and thus chemostatic selection for the maximum economy of reproduction. Thus, no r-K continuum, of which most contemporary textbooks of ecology speak with contempt (and whose purported existence meant that the concepts of r and K strategy ceased to be popular), but two quite distinct strategies functioning in nature, as was once assumed by the creators of the theory. End of the turn-off for my dear readers with good memories.

Let us finally return to invasive species and the hypothesis of the escape from natural enemies. If the size of the population of the invasive species in the original area of occurrence was regulated turbidostatically, e.g. by the activities of a parasite, then moving the population outside of the original area of occurrence could mean that the new population would escape from the reach of the particular parasite and thus from the effect of the given type of regulation. The equilibrium size of the new populations, determined under the new conditions, either chemostatically or turbidostatically through a different kind of parasite, can thus increase substantially and the species can begin to spread without control from the new territory to ever newer sites. Because of my present place of employment (Department of Parasitology), I am strongly inclined to favour all theories that attribute great importance to the role of parasites in natural processes. However, in my opinion, the theory of escaping from parasites and predators has a number of serious inadequacies.

It is especially not so easy to escape from parasites (and the theory attributed the greatest role to them). The more common, and these are certainly more important from the standpoint of regulation of the population, are brought along with the host species, or rapidly catch up with it. Parasites are frequently far better adapted for spreading to new sites than their hosts. They must have the ability to get out of the population of their hosts, which could well be leading a miserable existence because of them and which could thus very easily die out, to new sites where so far parasite-free populations live. As soon as an invasive species spreads to a greater territory, it is very improbable that it would not be immediately followed by its main parasite and that it would not immediately decimate its population to the original equilibrium value. In addition, as well as parasites narrowly specializing in a particular host species, there are also parasites specialized in "not specializing" and they concentrate on the most common species of potential hosts. As soon as an invasive species forms a sufficiently dense population at a new site, it directly asks for an originally "nonspecialist" to specialize on it.[7]

Renewed plasticity can be blamed for everything (what else!)

Thus, if we did not want to be satisfied with the theory of escape from parasites, what other possibilities are there? How about **escape from genetic variability and the consequent temporary return to the stage of evolutionary plasticity?** An evolutionarily plastic species would have a substantial advantage over its evolutionarily frozen competitors, as it could better adapt in new habitats to changes occurring in the particular environment since the time when it and the other species became evolutionarily frozen. Of course, because the limited number of individuals of the invasive species in the new territory would lose most of their genetic variability, they would become not only evolutionarily plastic, but also quite vulnerable to attack from parasitic organisms.

Once again, a small turn-off. A large number of biologists, and not only parasitologists, are of the opinion that the battle between parasites and their hosts play an important role in evolutionary processes. The **Red Queen hypothesis** even assumes that sexual reproduction was invented by evolution or at least is maintained in nature as an effective defence of hosts against parasites. If only because of its faster reproduction, a parasite is usually capable of faster evolution that its host. Consequently, it is usually capable of responding rapidly to evolutionary moves and counter-moves of its host and is capable of rapidly adapting to the traits of the local host population. If the host population is genetically diverse, it will adapt to the commonest host variant. The commonest variant is the one that exhibited the greatest biological fitness in the previous generation. However, this means that biological fitness actually exhibits negative heredity in the modern parasite-infested world. The variant that has high fitness multiplies and thus attracts the main attack of parasites and thus has lower fitness in the next generation. The core of the Red Queen hypothesis is the assumption that sexuality ensures that neither the genotype nor the phenotype is inherited from one generation to the next (see Chapter 8) and thus effectively protects the host population against parasites.[8] In a world full of parasites, the heritability of biological fitness is increased from **negative values to zero**.

If this explanation of the formation and maintenance of sexuality seems improbable to you, do not give up hope. This is only because I have not yet bothered you with other explanations. The development and maintenance of sexuality remains one of the greatest puzzles of evolution. There are a great many hypotheses for this. Many of them look, at first glance, to be far more sensible than the Red Queen hypothesis. However, on closer inspection, it is frequently found that they cannot function. The originally rather unusual Red Queen hypothesis has not yet encountered this fate. So who knows?

And now, back to invasive species. Study of the resistance to parasites in the old and new areas of occurrence and study of the reproductive system of invasive species could assist in deciding between the theory of escape from parasites and the theory of renewed evolutionary plasticity, as an explanation for the spreading of invasive species. If the theory of escape from parasites were valid, parasite-resistant sexually reproducing species should form the majority of invasive species (see again the Red Queen theory). In their new area of occurrence, invasive species should also be less attacked by parasites, on an average, than in their old area of occurrence. On the other hand, if the theory of renewed evolutionary plasticity were valid, asexual species should be more common amongst invasive species and, in its new area of occurrence, the invasive species should be attacked more by parasites than in its old area of occurrence (as their members have the same genotype and thus a parasite can adapt to them more readily).

Greater evolutionary plasticity of members of the invasive species in the new area than in the old area could, of course, be tested in experiments by exposing both populations to artificial selection. However, it should be borne in mind that, in these cases, it would be necessary to take into consideration and estimate in advance the probable primary acceleration of evolution in the genetically diverse population through the effect of the potential for selection from the alleles already present in the population (see Chapter 13). Thus, it would be necessary to begin to monitor the evolutionary response of the population to the relevant selection from a certain generation (say the 30th), or it would be necessary to study only the rate of evolutionary spreading of selected alleles, which would be introduced newly into both compared populations at a certain instant (i.e. populations derived from the original and from the new area of occurrence).

How to extinguish a flood

Which explanation of the formation of invasive species (of the two possibilities described above) is correct, could be of considerable practical importance in combating invasive species. If the hypothesis of escape from parasites were valid, then it would be necessary to concentrate on introduction of the relevant parasite into the population in the new area of occurrence of the species. On the other hand, the transfer and introduction of new individuals of the invasive species from the old to the new area of occurrence would make about as much sense as an attempt to extinguish a flood with water. Not only would this increase the number of individuals of the invasive species in the new location, but it would also increase the genetic variability within these populations and thus their resistance to any parasites. In case of the validity of the hypothesis of renewed evolutionary plasticity, on the other hand, it would be necessary to concentrate on introduction of the greatest number of genetically different members of the invasive species derived from the old area of occurrence into the population in the new area (preferably only males which would compete with the males already present and would thus not excessively accelerate the spreading of the invasive species). This introduction could increase the genetic variability of the population of the invasive species and thus eliminate its evolutionary plasticity. (This method would, of course, not work if the invasive species reproduced only asexually in the new area.) The transfer of parasites from the original area of occurrence could retard the spreading of the invasive species, but would probably not be capable of stopping the invasion and, in addition, would be connected with the risk of endangering the original local species by a new species of parasite. A parasite introduced by humans could thus even assist in spreading the invasive species, as it could be more of a danger to the local competitors, who have not yet encountered it and are not adapted to it.

Summary and incitement

In this chapter, we have shown that evolutionary plasticity of sexually reproducing species can have a positive effect on the functioning of ecosystems and particularly on the preservation of biodiversity. Asexual species have greater evolutionary plasticity. On the one hand, this allows them to occupy more extreme environments and utilize unusual natural resources; on the other hand, it endangers their capacity to peacefully coexist with other species. The ecological success of invasive species could be caused by the fact that, as a consequence of their isolation from the parent population and their low genetic variability (following from the small number of individuals in the population), they could return to the state of evolutionary plasticity and could thus gain a temporary advantage in the new area of occurrence over their evolutionarily frozen competitors. In the next chapter, we will show that the theory of frozen plasticity offers a very simple explanation for evolutionary trends – slow changes in the body structure of organisms taking tens of million years and thus exceeding the period of existence of the individual species. Before we get to this explanation, we will discuss the usual explanation of evolutionary trends, i.e. species selection and its possible product – sexual reproduction, and also the role of this sexual reproduction in the creation of new species.

Footnotes

1. Practically all studies state that the biodiversity of microorganisms is much smaller than the biodiversity of macroscopic organisms and that the same species of microorganisms occur throughout the world. However, this could be a result of lack of precision in our studies – it is possible that we are simply not capable of distinguishing between the individual species and thus erroneously assign them to the same species. As soon as we begin to use modern molecular taxonomic techniques, the individual species of microorganisms suddenly become differentiated into a large number of individual species.
2. Understandably, in actual fact, organisms do not directly form niches for other organisms, but only environments and resources. It probably makes sense in this connection to speak about "potential niches"; actual niches are formed only by organisms that have learned to use these resources.
3. Details about the family genetic tussles of this species of ant can be found in *Nature* 434: 1230–1234, 2005.
4. Newer summary articles on the subject of genome imprinting can be found, e.g. in *Early Human Development* 81: 73–77, 2005, *Reproduction* 122: 185–193, 2001; this subject is discussed in a very nice and comprehensible form in the book Ridley M. *Mendel's Demon: Gene Justice and the Complexity of Life.* Weitenfeld & Nicolson (London), 2000.
5. The classical (and, in my opinion, inadequate) explanations of the paradox of invasive species, i.e. their ability to prevail in competition over the locally adapted species, is described, e.g., in *Global Ecology & Biogeography* 9: 363–371, 2000.
6. A description of turbidostatic and chemostatic population regulation and the ecological and evolutionary consequences of both types of regulation can be found in the *Journal of Theoretical Biology* 188: 121–126, 1997.
7. A comparative study on two species of freshwater fish has shown, for example, that the diversity of parasitic fauna has a tendency to increase in introduced species, see *Journal of Biogeography* 30: 837–845, 2003. There are, understandably, also a number of studies that, on the other hand (in accordance with the expectations following from the original theory of escape from enemies) indicate a lower level of parasitic infestation amongst introduced species, see, e.g., the review article in *Nature* 421: 628–630, 2003. It is not entirely clear when one or the other case occurs. A problem also lies in the fact that, in a great many works, there is no strict differentiation between species introduced to a foreign territory and species introduced to a foreign territory that act as invasive species. In my opinion, the aspect of the level of parasitic infection of invasive species requires further study.
8. I can highly recommend the popular book by M. Ridley *The Red Queen.* Or you can go directly to the source and read the book by G. Bell *The Masterpiece of Nature: The Evolution and Genetics of Sexuality.* University of California Press (Berkeley), 1982. A very nice review article on this subject can be found in *Proceedings of the National Academy of Science, U.S.A.* 87: 3566–3573, 1990.

CHAPTER 16 Could the theory of evolutionary plasticity explain the existence of evolutionary trends?

Things work out differently in the world of frozen species

There are a number of important differences between the Darwinist system of evolutionarily plastic species and frozen world of evolutionarily elastic species. In the Darwinist world, the occurrence of mutually different and sharply differentiated species is actually only a side product of our temporally limited view of nature. The fact that we usually study biodiversity only on a certain time plane means that we think there are sharp boundaries between the individual species. Over a longer time interval, the individual species gradually change, smoothly merge into one another, mutually differ or, following recombination of the individual populations, merge into a single species. In the world of frozen plasticity, each sexually reproducing species is formed as a consequence of a unique event and passes through two periods during its existence, a short period of evolutionary plasticity, during which it can change as a consequence of selection pressures, and a substantially longer period of frozen plasticity, during which it changes very little and most of the changes that might still occur during this period are reversible. Because of the disproportion in the duration of these two periods, most of the species that we encounter in nature at a particular moment must necessarily be in the stage of frozen plasticity and, amongst other things, cannot adjust to changing conditions.

This fact has a number of important consequences. For example, the number of evolutionary changes that occurred in a certain group of organisms should not depend directly on the time that has expired since this group began to develop from a common ancestor, but rather on the number of speciation events that happened to this species during this time. Of course, the number of speciation events and the time of development of a certain group of organisms are usually quite closely connected. However, modern statistical methods allow the two effects to be separated and to determine whether the actual data correspond better to the Darwinian model of constant plasticity of the species or rather to the world of evolutionarily frozen species. Such a study has recently been performed on songbirds (passerine birds – *Passeriformes*) and its results indicate that, at the very least for this group, the number of evolutionary changes corresponds to the number of speciation events rather than the lifespan of the relevant evolutionary branch. It thus follows that the evolution of songbirds corresponds better to the model of frozen plasticity than the gradualistic model of Darwinian evolution.[1] The model of frozen evolution also allows us to explain the existence of evolutionary trends.

Where do evolutionary trends come from?

Palaeontologists, i.e. scientists concerned with the actual history of evolution on the Earth (usually through the study of fossils) and evolutionary biologists, i.e. scientists concerned with the mechanisms of evolutionary changes, do not always see eye to eye and do not always communicate sufficiently. This was, amongst other things, the reason why the theory of punctuated equilibrium had to wait so long before it was discovered. Evolutionary biologists simply did not realize that there was a basic discrepancy between their theories and palaeontological data. Another similarly interesting phenomenon that is well known to palaeontologists and that is not known to or is underestimated by evolutionary biologists consists of **evolutionary trends.**[2] An evolutionary trend is a change in one of the traits of organisms that occurs in a certain evolutionary lineage over long periods of time. These periods are several times longer than the average duration of individual species.

The thing that is interesting and incomprehensible about evolutionary trends is their slowness. At first glance, it would seem that a trend can be as slow as it wants. This is not the case. If a particular adaptive trait is to be formed through the action of natural selection, then its evolutionary development cannot last too long. If, for example, over 20 million years, the bodily proportions of the representatives of a certain evolutionary branch increase at a constant rate from 10 cm to 2 metres, we would expect an average increase in size of about 1 mm over 10 000 years. If the average generation time of the particular organism were 1 year, then this would correspond to an average change of 0.1 mm per thousand generations. However, this is a difference in size that is quite invisible from the standpoint of natural selection. It can be derived mathematically that, in order for natural selection to be able to decide on the fate of a particular mutation, rather than the random processes of genetic drift and draft, then the selection advantage of the particular change must exceed a certain minimal value. Thus, the evolution of ten-centimetre organisms can certainly not proceed at the snail's pace of 0.1 mm per thousand generations. How can we escape from this conundrum?

A number of solutions have been proposed. The classical solution of the problem assumes that the trend occurs as a fast evolutionary response to an extraordinarily slow change in the environment. For example, to a slow change in the amount of oxygen in the atmosphere, to lengthening of the day as a consequence of slowing down of the rotation of the Earth, etc. Other theories assume that evolutionary trends are not caused by natural selection, but by other evolutionary processes. This could be the **wall effect**, which was mentioned in connection with the formation of complexity in Chapter 5. If the size of the members of a certain evolutionary line was limited from the bottom at the instant of its formation, e.g. by a minimum number of cells that are still capable of ensuring viability of the particular type of organism, then, as a result of the wall effect, the organisms will move away from this value during further evolution and will gradually get larger at an arbitrary (although probably decreasing) rate.

Another, basically quite similar explanation of evolutionary trends assumes the action of developmental constraints. Proponents of the theory of developmental constraints assume that, because of the nature of specific mechanisms controlling developmental processes in a particular group of organisms (ontogenesis), randomly occurring mutations to the phenotype of the organism will have a much greater tendency to change the phenotype in one direction rather than the other. In one evolutionary line, the effect of random mutations can thus favour the tendency to increase body dimensions (or, for example, decrease the number of digits on the limbs). However, the trends can be quite the opposite in another line. At first sight, the explanation based on the action of developmental constraints might seem complicated and hard to understand. However, an illustrative example that was first introduced into evolutionary biology by Charles Darwin's half-cousin, Sir Francis Galton, might make it easier to understand. Imagine a bead on a smooth horizontal surface. If we gradually act on it with various forces from various directions, then the direction and the distance through which the bead travels will depend only on the direction and force of our action. However, if we place an irregular polyhedron on the surface instead of the bead, then the character of the response to our action will be determined to a considerable degree by the shape of this polyhedron. In a certain direction, the polyhedron will move (roll) very easily, while it will be very unwilling to roll in other directions. If our action is random in direction and force (analogous to random mutation), then the direction of movement of the polyhedron (the direction of evolution of actual organisms with specific developmental mechanisms, i.e. mechanisms of transfer of genotype to phenotype) will be decided by the shape of the polyhedron – developmental constraints. Similar laws are certainly valid in the development of organisms. If the patterns on the wings of butterflies are formed by a substance (morphogen) diffusing from a certain point in all directions, determining the formation of a certain colour, then circular patterns (eye spots), rather than, say, squares or triangles, will have a greater tendency to form on the wings of all newly forming species of butterflies.

The many faces of a selection

A further explanation of evolutionary trends assumes that the driving force for the formation of a great many trends is **species selection**. Before we get to explanation of the hypothesis of the formation of trends through the action of species selection, it will be useful to recall the differences between the concepts of Darwinist individual selection within a species, selection between species, and species selection. According to Darwin, the driving force for evolution lies within a species, i.e. **individual selection**. The basic unit for individual selection is the individual who competes with other individuals within the population of its own species as to who will produce more progeny for the next generation. Nonprofessionals have a tendency to confuse **selection between the species** with Darwinian individual selection. In inter-species selection, the individual species compete as to which of them is more capable of utilizing the common resources, i.e., more effectively converting the available resources to progeny under the given conditions. Inter-species selection is frequently used to explain a situation where some traits seem to be advantageous for the species as a whole, but disadvantageous for their carriers. A typical example is the ability of birds to reduce their own reproduction (number of eggs laid) when the size of the population at a particular site begins to approach the survival limit for the particular environment. In the vast majority of cases, inter-species selection is not the correct explanation of the formation of this type of **altruistic traits**, i.e. traits that provide one's own species with an advantage compared to other species, but also place the carriers of the particular altruistic trait at a disadvantage compared to the members of the same species that do not bear this trait. On an average, altruistic individuals leave fewer progeny than their more selfish competitors and consequently their genes causing altruistic behaviour are gradually forced out of the population by individual natural selection. A species consisting of altruistic individuals gets along better than a species consisting of selfish individuals (it utilizes the resources in its environment more effectively and thus achieves a greater number of individuals in the population on the long-term average), but mutants – selfish individuals – are still best off.

At the present time, the possibility of the action of inter-species selection (and its related **group selection**, in which individual populations within a single species compete) has not been completely rejected with such vehemence and certainty, as was still the case in the 1980s. Mathematical models have shown that, under certain conditions, both types of natural selection (inter-species and group) can occur. Nonetheless, it is almost certain that most altruistic traits were fixed during evolution by some form of individual selection. In the above case, individuals that stopped reproducing under conditions where the density of the population approached a certain limit were only apparently altruists. In actual fact, most of their progeny would have died anyway, from hunger or from the action of parasites, so that, from the standpoint of biological fitness and in the given situation, it was preferable to reduce reproduction in time and save strength for the time when conditions improve, perhaps in the next reproductive season. Another aspect of inter-species and inter-population (group) selection will be discussed further in Chapter 18, where we will show that the theory of frozen plasticity presents a more optimistic view of these types of natural selection than the Darwinian or Dawkinian theories of evolution.

How the weakling won out over the super-mouse (in species selection)

So we have finally got around to species selection. In **species selection**, the individual species do not compete to see which of them is capable of more effectively utilizing common resources, but the individual evolutionary branches compete as to in which speciation will occur more frequently and in which extinction will occur less frequently. Imagine that we have two species of mice. One is a super-mouse that is capable of living in practically any environment, can eat practically anything, and is easily able to overcome any natural obstacle (mountains, rivers, even oceans). The other is a weakling mouse that can survive only in the lowlands, only where the wind is not too strong and there are no sharp stones that would prick its sensitive little feet. While the super-mouse will easily occupy all the habitats and create one large interconnected super-population, the weakling mouse will form a large number of small local populations, between which exchange of migrants will occur only very occasionally. Which of the two species will split off more daughter species over time? The weakling, of course. Most of the local populations will disappear quite quickly, what else could you expect of weaklings; however, because of their geographic and thus genetic isolation, a number of them will result in the formation of a new species. And sooner or later, the right virus will appear amongst the super-mice and cause its entire population to die out. In fact, such a virus will appear amongst the super-mice, which form a single large global population, far sooner than amongst the species of mice forming only small populations.[3] A virus can attack anywhere and the subsequent pandemic will affect the entire population. While the evolutionary branch of weaklings will manage to repeatedly branch out by that time and the extinction of one or two species would not mean anything particularly awful for it, the branch of super-mice remained unbranched and would become completely extinct with the extinction of its single species. This scenario will undoubtedly happen to the evolutionary branch of anthropoids, unless it takes its fate into its own hands and thus saves for the future the Abominable Snowman, a weakling that, similar to Snow White[4], apparently melts in the warmth and thus its isolated small population remains only on the slopes of some high mountain ranges. Let us hope that, by that time, things do not get too warm because of the greenhouse effect (the Snowman would melt) or that it does not get too cold (he might carelessly come down to the foothills where he would gain voting rights, AIDS and bird flu). Another reason to stop releasing greenhouse gases: how would our successors, the lords of the planet, the wise octopus *Octopus sapiens* (or perhaps the wise rat *Rattus sapiens*) manage, if they could not go to the zoological gardens on Sunday with their families to look at the last members of the formerly evolutionarily very successful branch of anthropoids?

Actually, species selection has the last word in evolutionary processes. A super-mouse might be a hundred times better in direct competition for resources; but if it does not keep pace with its competitors in species selection, then the weaklings will finally triumph. Then why do we see so many traces of classical selection in nature (competition for the greatest biological fitness), i.e. so many adaptive traits increasing the effectiveness of utilization of resources and, simultaneously, so few traces of species selection (competition for the speed of speciation and the slowness of dying out)? For two reasons. To begin with, species selection has a rather narrow area in which it can be effective. The vast majority of traits that increase the probability of speciation simultaneously increase the probability of extinction. For example, species occupying large territories die out more slowly but, simultaneously, it is harder for new species to split off. So that, very frequently, the advantages and disadvantages tend to cancel out.[5]

The second reason is less obvious, but is apparently more important. Conspicuous adaptive traits created over the progress of evolution by classical selection, such as the chamber eye or the placenta, are almost always formed by gradual accumulation of small evolutionary changes as a consequence of individual mutations. Species selection does not have enough time to create complicated adaptations leading to greater speciation rates in a similar manner. As we mentioned in the chapter concerned with the formation of biodiversity, the duration of life on Earth is too short compared to the average duration of survival of a single species. While 5 million generations of an animal with a generation time of one year fit into the average time of survival of a species, of the order of 100–200 successive species in one evolutionary line[6] would fit into the period of duration of multicellular life on this planet, i.e. into a period lasting, say, 700 million years. The method of trial and error, on which evolution is dependent, cannot be tried out very well in such a few "generations". In addition, the number of species in a particular evolutionary line that occur simultaneously at any instant on the Earth and that compete in the discipline of the rate of speciation is many orders of magnitude smaller than the usual number of individuals – members of a single species.[7] While species selection could have the final word in evolution, no miracles can be expected from it because of its incomparably smaller effectiveness. Especially not the formation of complicated adaptations requiring the accumulation of a great many independent mutations, intended to increase the probability of speciation. There was not enough time and there was not enough "experimental material".[8]

Where did sex come from?

By the way, it cannot be completely excluded that at least one complicated adaptation increasing the speciation rate in species selection was actually formed and thus baffled future evolutionary biologists. This adaptation could consist of sexual reproduction and the relevant molecular apparatus required to provide for it. As I mentioned above, the formation of sexual reproduction is the greatest puzzle of evolutionary biology. It is not difficult to find a great many very good reasons why sexual reproduction would not be formed or, if it were formed in a species through a strange interplay of evolutionary fates, why it would rapidly disappear again as a consequence of competition with asexually reproducing mutants. Of all these reasons, I will mention only the two most important here. These are the twofold cost of males and the twofold cost of meiosis.

The **two-fold cost of males** is easier to explain. A population that would consist of only asexually reproducing females would reproduce at twice the rate of a population of sexually reproducing species that sinks half of its resources in the production of „nonreproducing" males. It can, of course, be objected that the population would not have to be half males, that one male in the local population is frequently sufficient for the purpose of reproduction, so that the overall loss in the rate of reproduction need not be so great. The problem lies in the fact that this arrangement is evolutionarily unstable. Under these circumstances, a single male in the population would become the father of all the offspring and would thus transfer a greater number of his genes to the next generation. The females would immediately start to compete for the position of mother of the future male. And consequently the number of males in the progeny would very rapidly return to the usual value of 50%. It is obvious that the two-fold cost of males does not apply to all species occurring in nature. This is not true, for example, amongst hermaphrodites, i.e. organisms in which one individual plays the role of both the male and the female (forms both types of sex cells). And it is also not fully valid in species in which both parents participate to a comparable degree in formation of the embryo or subsequent care for the progeny. Thus, it is not true for a large number of protozoa, in which the zygote is formed by the combination of two comparably large cells, or in most birds and some mammals, where both the mother and father invest resources into care for the offspring. Nonetheless, there are an enormous number of species where the two-fold cost of males is valid that, in spite of this clear disadvantage, did not succumb to competition from asexually reproducing mutants.

The **two-fold cost of meiosis** is valid for all sexually reproducing species. It consists in the fact that, in meiosis, the female basically throws away half her alleles (forces them out to the polar bodies, which are destroyed – see Box 6.5 on p. 89) and thus allows them to be replaced in the future zygote by foreign copies of the relevant genes of the male. If a mutated gene were to appear in the genome of the female that would prevent this wasting of its own genetic material, for example, that would cause that the female germ cell to retain both sets of chromosomes and, following fertilization by the male sex cell, would destroy the chromosomes derived from the male, it would be passed on to the next generation in twice the number of copies compared to the original unmutated allele. A number of mutually unrelated asexually reproducing species do just this, so that, technically, it is probably not a difficult matter and the relevant mutations occur in nature relatively easily and quite frequently. Once again, we would expect that asexually reproducing mutants should spread rapidly in the population at the expense of their orderly sexually reproducing competitors and that, in a few generations, the species would change from sexual reproduction to asexual reproduction.

For about 50 years, evolutionary biologists have unsuccessfully sought for the reasons why this does not happen in nature and why the vast majority of species so tenaciously retains this apparently disadvantageous means of reproduction. The solution that immediately suggests itself, i.e. that organisms like to reproduce sexually because they enjoy it can be readily ignored. Evolution is not in the least interested in what organisms enjoy or do not enjoy, only in what is advantageous from the standpoint of their survival and reproduction (in actual fact, rather that which is evolutionarily stable, see Chapter 7). To be more exact, evolution itself decides what is

Box 16.1 Rabbit droppings

I must apologize for readers with weak stomachs for the rather harsh example of rabbit droppings. Instead of this, I could have said June bugs or ant pupae; however, in this case, I would not be able to mention that rabbits (and a great many other herbivores) actually do eat their own droppings. They basically process their hard-to-digest food twice because they do not have a specialized system of several stomachs like ruminants. Instead, they simply let the food go through their digestive system twice. The professional literature does not mention whether the special form of droppings, which they place in a separate location before consuming it again, taste like walnuts (to them) or not.

going to be pleasant or unpleasant for us. If it were advantageous for evolution for us to eat rabbit droppings (for example, if they contained some rare vitamin) then evolution would certainly have given us taste buds that would tell us that rabbit droppings taste like walnuts.

Evolutionary biologists have written thousands of professional articles on the subject of the formation of sexuality and the more famous of them have even written books on this subject. (Perhaps I will also write one when I get to be sufficiently famous; books with the word sex in the title certainly sell well.) Several dozen potential reasons have been proposed, from the need for sexuality to repair mutations in the genome, through acceleration of micro-evolutionary processes, through retention of genetic variability, or the formation of unusually capable recombinants (a genetic elite), to the above-mentioned defense against rapidly evolving parasites through limited phenotype inheritability, and thus an increase in the biological fitness from negative values to zero (see Chapter 15). Nonetheless, it still seems that we have not found the proper solution. (It is quite probable that authors of the individual hypotheses will think up something else. However, if there are 30 of these hypotheses, at least 29 of them are probably erroneous.)

The thirty-first (and certainly finally the correct) hypothesis for the formation of sexuality

What if we have been looking for the right solution in completely the wrong place? What if sexual reproduction actually does reduce the biological fitness of its carriers and reduces their chances in individual, group and inter-species competition? What if the product of species selection, adaptation to the most frequent speciation, is involved? It would be logical. Sexual reproduction is the thing that holds most species together, it means that they do not disintegrate into a tangled ball of more or less mutually related and more or less similar lines, which compete in the given environment and are capable of completely suppressing one another in competition. In sexually reproducing species, the spreading of a newly formed advantageous mutation is not connected with the disappearance of the other lines, as sexual reproduction means that the advantageous mutation will gradually move into the genome of its competitors (or, more exactly, their progeny). In asexually reproducing species, the spreading of a new advantageous mutation is accompanied by the disappearance of the other lines.

In addition, sexual reproduction is a very effective motor for the actual process of speciation. For example, several hundreds of species of cichlid fish, which were formed in large African lakes, probably over the past 100 000 years[9], could tell us interesting tales. Speciation very frequently occurs in that a group of individuals within a certain species begins to reproduce preferentially or even exclusively amongst themselves. For example, some females begin to prefer, for example, males with a red spot on their tails. Not that these males would differ from other males in any other way; it's just that, if a female wants to be "in", then she certainly canot throw herself away with a male with a yellow stripe on its tail. Personal taste is of no importance, fashion is fashion. Males with a red spot suddenly have a great advantage in the competition with males with a yellow stripe and consequently leave behind more progeny on an average. Thus, the proportion of individuals with a red spot understandably increases. However, what is not so apparent at first glance is that the proportion of females preferring males with a red spot also increases. A randomly selected male with a red spot carries on his genome not only the gene for the red spot, but also a gene for preferring males with a red spot. Why? Why, because his father probably also had a red spot and his mother, when she chose this father, probably carried a gene for preferring males with a red spot. Thus, if the gene programs the female to preferentially mate with a male with a red spot, then it helps to spread not only genes for the red spot, but also copies of itself, i.e. spreading of the gene for preferring a red spot.

Can you still remember Dawkin's model of Green Beards in Box 8.5 on p. 115? To make sure, I will remind you. This was a hypothetical gene that gave its carriers green beards and also made him prefer other, quite unrelated, bearers of green beards. Dawkins was of the opinion that the occurrence of such a gene in nature is possible, but is not very probable. A single gene or genes located close together on the chromosome would have to control two different, very dissimilar phenotype traits, green beards and preference for green beards. As follows from the example above of preference for males with a red spot, the situation for implementation of the green-beard mechanism is much more favourable. The genes for green beards (red spot) and for preferring green beards (red spots) could each be located on a completely different chromosome. Nonetheless, as soon as a female helps a male with a green beard (enables him to reproduce), she automatically increases the proportion of genes for preferring green beards. The famous Freiherrn von Münchhausen of renowned memory saved himself and his horse when he pulled himself out of the bog by the hair. I am by nature a trusting person (well, actually, I'm not, but I'm not giving up a catchy phrase for that reason), but somehow I suspect that this would not work in practice. (But I will probably not try it out in practice. I would probably manage to find myself a horse and a suitable bog, but my hair has recently been disappearing at a disturbing rate and I might just pull the last strands out before I disappear in the mud.) Nonetheless, this actually works in a certain sense in the world of genes; a mutated gene for preferring a certain trait pulls itself out of the gene pool all the way to fixation. Consequently, I once proposed the term **autoelevation** for this phenomenon, which was described in the 1930s by the famous statistician and evolutionary biologist Sir Ronald Aylmer Fisher (there are never enough terms). The name has not caught on so far; I am afraid that it is used only by students arriving at my office for an examination in evolutionary biology. If I have enough students, then, in time, who knows? So, to make certain, try to remember it.[10]

If the preferred traits are not mutually exclusive, for example a red spot on the tail and a green circle around the eye then, as a consequence of autoelevation (it is necessary to repeat the new term), the patterns on the body of the particular species can gradually become more complicated. Maybe this is the reason why nature is so colourful and Portmann and other aesthetics of nature have something to think about.

Box 16.2 Adolf Portmann (1897–1982)

Portmann was a Swiss biologist who studied the biological meanings of the external appearance of organisms in the first half of the 20th century. He showed that there are a large number of conspicuous structures, patterns and colours on the surfaces of the bodies of animals, which were very frequently formed as a means of communication within the species and between species (called address phenomena). However, in addition to address phenomena, nonaddress phenomena occur in nature, which Portmann considered to be manifestations of a general tendency of all organisms towards self-presentation. In my opinion, a great many of these structures were formed as a consequence of autoelevation and, in organisms that are not equipped with sight, as a form of warning or masking coloration. From the standpoint of spreading any newly formed alleles, it is certainly advantageous if they increase the viability of their bearers. However, it is even more advantageous if there is an external indication of their presence in the genome (such as a red spot on the tail), which opens the possibility of very rapid spreading through autoelevation. After the Second World War, German biology (and all works written in German) were completely pushed aside by American biology, as a consequence of which the aspects introduced, e.g., by Portmann, remained practically unknown to modern biologists who are thus not greatly affected by these works.[11]

If the new traits are mutually exclusive, like a red spot on the tail, blue spot on the tail, red circle on the tail, autoelevation (repetition is the mother of wisdom) can lead to the formation of several species. Initially, their members can differ only in the preferred trait and in preference for the trait; however, after a sufficiently long period of time, incompatible alleles will necessarily accumulate in their genomes and the members of the newly formed species will thus lose the ability to reproduce together (see the model of the formation of a reproductive barrier through the accumulation of mutually incompatible alleles on p. 85). A large number of species of African cichlids have not yet reached this stage, to their detriment and ours. Today, when we have nicely polluted their lakes with organic substances (I could say eutrophized, but I do not want to overdo it with the professional jargon) and, as a consequence of the resultant multiplication of algae, the clarity of the water in the lakes has been reduced, the females of a great many species of cichlids can no longer recognize the males of their species on the basis of body patterns. Thus, they reproduce more or less blindly, which is still easy enough, but unfortunately interspecies crossing means that the formerly clearly distinguishable species are merging together and their differences will disappear.

Sexual reproduction increases the probability of the formation of a new species not only by the mechanism of autoelevation (that was the last time, a well established term does not need to be repeated) and a number of other mechanisms. Some of them are quite interesting, for example **reinforcement**; however, because this chapter is getting disproportionately long and we have not gotten to the core of the matter because of the constant turn-offs from turn-offs from turn-offs, I will not spend more time on it here. Buy yourself a textbook on evolutionary biology (preferently, of course, my textbook), everything is there. I hope I can conclude that sexual reproduction basically increases the rate of speciation and could quite readily be the above described complicated product of species selection.

It's easy for large species to speciate, isn't it?

Let's return to species selection as one of the possible explanations of the formation of evolutionary trends. Because of its low efficiency and the slowness of the evolutionary processes that are dependent on it, **species selection** could be the motor for long-term evolutionary trends. For example, imagine that, in some species of animals, Darwinian selection could favour an increase in body dimensions, while it favours a reduction in other species. However if, on an average, large animals exhibit a greater tendency for a new species to split off, species selection over long periods of time will cause an increase in the body dimensions of species within a particular evolutionary line. And why should large species have a greater tendency to split off new species than small species? Why not, that was just an example. However, because a trend leading to an increase in body dimensions within an evolutionary line is very frequent in nature, in fact so frequent that it has its own name, **Cope's rule**, I will provide a possible and quite probable explanation free of charge. In any case, it's not mine, at least as far as I know, it was proposed by S.J. Gould himself. Large animals necessarily have smaller populations than small animals. And a small population more readily evolves into a new species than a large population, if only because its members more readily overcome valleys in the adaptive landscape, see Chapter 15. Or because it can be very readily reduced to a very small population and can thus pass from the state of evolutionarily freezing to the state of evolutionary plasticity. And then it is, of course, easy to speciate.

Trends in the world of freezing species

This is thus the basic known explanation for evolutionary trends. And now, finally, a new one, towards which I was working the whole time. It is based on the usual Darwinian individual selection acting with sufficient, i.e. considerable, intensity on the species of a certain evolutionary line and on the effects of the phenomenon of evolutionary freezing. If a species is capable of responding to selection pressures only for a small part of its existence, the resultant rate of the evolutionary process is small even if the relevant selection pressure is quite strong and if it acts uninterruptedly over a long period of time. As a consequence of the relevant selection pressure, the newly formed species sets off in a certain direction, say towards an increase in body dimensions; however, before it gets very far, before it gets to its optimum size, it evolutionarily freezes. And evolution of body size must wait, say, a million years until further speciation occurs and it can go a bit further.[12]

So you can see, it is quite simple. It would probably not be enough to fill a whole chapter. So you should not be too surprised at the turn-offs and turn-offs from turn-offs.

Summary and incitement

Although it might not be entirely clear after so many turn-offs from turn-offs, the core of this chapter should consist of two statements. These are the statement that, according to the theory of frozen plasticity, the size of evolutionary changes should reflect the number of speciations in a particular evolutionary line and not the duration of the line, and the statement that the theory of frozen plasticity offers a new solution to the puzzle of the existence of slow evolutionary trends, long-term changes in the phenotypes of organisms whose duration frequently exceeds the usual time of existence of the species by many times. Evolutionary trends could be so slow because the species can respond to selection pressures only for a short part of their existence, i.e. immediately after their formation. Then they evolutionarily freeze and must wait for further speciation before they can evolve further. I will not tell you in advance what is contained in the next chapter. At least at the theoretical level, I favour the idea that it is useful from time to time to break away from the established routine and try something else. Perhaps I will just mention that it will be related to species bred in captivity and domesticated species.

Footnotes

1. Comparative studies documenting the role of the number of speciations in the development of the morphology of song birds were published in *Nature* 438: 338–441, 2004. Molecular study showing the same results was published in *Science* 314: 119-121, 2006.
2. The development occurring for internal reasons in a particular direction (thus exhibiting a certain spontaneous trend) is designated by the old term orthogenesis. This term was made popular by the German zoologist G.H.T. Eimer (1843–1898). Most hypotheses explaining orthogenesis rely on the action of known natural forces. The best known (especially in humanitarian fields) are, however, hypotheses that are based on the action of so-far unknown forces or tendencies, for example the purported tendency of living creatures (or even the universe) to improve themselves.
3. The theory of viral extinction assumes that a large part (and perhaps the vast majority) of species extinction outside of the periods of mass extinction caused by catastrophic climatic changes (impacts of small planets, flood volcanism) is caused by viruses. According to this theory, species forming a large continuous population are more sensitive to extinction caused by viruses than species forming a large number of mutually isolated populations. The theory is discussed on the basis of specific data in *BioSystems* 31: 155–159, 1993 and *Trends in Ecology and Evolution* 8: 209–213, 1993.
4. Here, I do not mean the Snow White who lived in a temporary polygamous relationship with seven dwarfs and later in a happy marriage with a somewhat necrophilic prince, but an individual of the same name, a girl that an old man and an old woman built of snow and who they forgot to tell that she must not jump over the camp fire with the other children. (Typical example of neglect of proper care!)
5. There is a trait that reduces the probability of extinction without simultaneously reducing the probability of speciation. This trait is the ability to actively fly. Species with this ability can have relatively small populations (which can increase the probability of speciation) and can simultaneously occupy large areas (and in their framework, move to a suitable area if necessary), which reduces the risk of extinction. It will probably not be accidental that both bats and birds (and also, in certain sense, "flying" fish) form taxons with an unusually large number of species. Don't search for the reference – this is a reference!
6. It is not clear how long multicellular forms of life have been present on our planet. Palaeontological data suggest 700 million years; in contrast, molecular phylogenetics yield older estimates (usually more than a billion years). As I am quite well acquainted with the problems associated with calibration of molecular clocks used to date individual evolutionary events by molecular phylogeneticists (and I am not well acquainted with similar problems that are probably encountered by palaeontologists), I have a greater tendency to believe palaeontological data. More species than the calculated 100–200 would fit into this period of time, as a species could speciate long before it becomes extinct; however, I doubt that it could be ten times more. This is also based on the fact that the splitting off of a new species apparently increases the probability of extinction, see *Paleobiology* 24: 305–335, 1998.
7. The numbers of members of an individual species understandably differs greatly even within a single taxon. For example, it is estimated that there are 100 billion birds on the surface of the Earth. This would mean that about 10 million individuals would correspond to each of the approximately 6000–8000 known species of birds. However, the vast majority of species will consist of fewer individuals by several orders of number, while the most numerous species will have several orders more. Nonetheless, I dare to state that the number of mutually

competing species is incomparably less than the number of individuals competing within a species. Only species that can meet in the same territory and that simultaneously have similar niches can compete. However, for each species, this will be only a tiny fraction of the total number of several dozen million species occurring on the Earth at the present time.

8. As one of the reviewers of the book, P. Baum, quite correctly pointed out, the situation is rather more complicated in connection with the number of experiments that macroevolution had available. Macroevolution could perform a great many experiments simultaneously – split off individual populations that could either develop into a successful species or could also unsuccessfully die out. Nonetheless, I insist that the number of experiments that microevolution has available during the existence of a single species (number of mutants) is incomparably greater than the number of experiments that macroevolution has available (number of started speciation events). This note would probably belong more in Chapter 4, as it is related to the formation of traits through the mechanism of species selection and also natural selection. However, without prior explanation of the theory of frozen plasticity, the reader could not fully appreciate the fact that one species corresponds to one period of evolutionary thawing, i.e. one chance for fundamental evolutionary innovation in the relevant evolutionary line. By the way, in this respect, Baum proposes a very interesting new definition of a biological species (valid for sexually reproducing organisms): "A species is a set of individuals sharing an identical gene pool in the time between two periods of evolutionary plasticity."
9. For quite some time, it was even assumed that all the 500 species of cichlids in Lake Victoria developed over 12 000–14 000 years, because the lake was completely dry before this time. However, the newest results indicated that a number of species developed in a different lake and moved to Lake Victoria only later, see *Science* 300: 325–328, 2003.
10. In his famous book, *The Genetical Theory of Natural Selection*, Dover Publication (New York), 1958, R.A. Fisher discusses two evolutionary phenomena simultaneously – the evolution of genes for male preference (autoelevation) and evolution of preferred male traits by the mechanism of run-away selection. (The hypothesis of run-away selection proposes that the development of male secondary sex traits, such as the peacock tail, could be caused by the fact that females always prefer the individual with the most highly developed sex traits from amongst the available males. As a consequence, the size of the particular trait increases over time, sometimes to such a degree that it reduces the viability of their carriers.) The fact that he did not sufficiently distinguish and separately name these two phenomena meant that many evolutionary biologists used the same name – run-away selection for both unrelated phenomena (without realizing it).
11. Sometime at the beginning of the 1990s, Daniel Frynta came up with the idea that large populations are able to fix primarily alleles that are externally manifested and thus utilize the Green Beard principle (during one of our periodical brain stormings – at that time, we did not have to write grant proposals and surveys of publishing activity and had time to think about scientific problems). It was my idea that the Green Beard model can readily function because of autoelevation for all genes that are externally manifested (and not only for those that program their carriers to assist other Green Beards). We never published our joint explanation of nature's aesthetics because of lack of time (and general laziness). Portmann's opinions are outlined, e.g., in the books *Neue Wege der Biologie.* R. Pipper (München), 1960, *Animal camouflage.* The University of Michigan Press (New York), 1959, or even better in the books by Z. Neubauer, e.g. *Biomoc.* Malvern (Prague), 2002.
12. During the preparation of the English version of this book I found an interesting paper written by Mark Webster: *Science* 317: 499–522, 2007. The paper shows that stratigraphically old and/or phylogenetically basal taxa (of trilobites) are significantly more variable than younger and/or more derived taxa. If this is a general trend, it can be possibly explained by slow continuous freezing of body planes in the course of macroevolution. The freezing of a body plane (as well as freezing of species) is probably not a yes/no phenomenon, but rather some kind of a continuum. During the macroevolution (and also during the freezing of a species) more and more traits change from plastic to elastic stage. Some traits can melt during speciation and the number of such melted traits negatively correlates with the number of founders of a new species. However, some traits are probably frozen forever because a large number of independent mutations would be necessary to occur in the same time to get the species out of the "genetic trap" based on networks of epistatic interactions. The freezing of body planes could in fact explain also the existence of the unique phenomenon of Cambrian explosion – a rapid origin of most major groups of multicellular animals in the fossil record, around 530 million years ago. A new radically different body plan cannot develop in later phases of macroevolution because all key traits of modern organisms are already permanently frozen.

Chapter 17 Sitting in the pub with the good and bad species of Daniel Frynta

My friend and colleague Daniel Frynta thoroughly enjoys breeding mice, snakes and lizards. There are certainly a number of these afflicted people amongst us; however, my friend Daniel is also a zoologist and ethologist, so he is quite capable of justifying this interesting psychological aberration to himself and, to a certain degree, also to his surroundings. So he has about ten rooms at home full of terrariums and every once in a while carries home another scaly little friend. While ordinary animal fanciers obtain primarily aesthetic and emotional satisfaction from their pets, for Daniel his charges are primarily a source of intellectual inspiration. A good scientist is characterized by the ability to answer questions that his colleagues cannot answer. An excellent scientist, like Daniel, can be recognized by the fact that he is even capable of posing questions that none of his colleagues have so far posed.

It is not long ago that, in a restaurant located not far from our faculty, over a half-litre of lager beer (which is the right amount for commencing the most fruitful discussion on the functioning of nature; the quality of the conversation deteriorates rapidly after the fourth one), Daniel put forward a problem that had been lying in his mind for some time. "It is strange and it could be important that some species of animals reproduce without problems when kept in captivity, and others do not, but rather die with the slightest excuse. Simultaneously, these could be two closely related species with very similar ecological requirements." He immediately suggested several possible solutions that could be tested experimentally (preferably on new mice and snakes that he would purchase for breeding especially for this purpose). The first possibility could be "**bad genes**". In order for the members of a particular species to reproduce well in captivity, a good breeding group must meet through the chance hand of fate. The males and females or, more precisely their genes, must fit together properly; otherwise the animals will also try to reproduce (what else to do all day long in a terrarium) but the offspring will have poor genotypes and will die, as mentioned above, at the first suitable excuse. However, the problem lies in the fact that the entire phenomenon is regularly repeated without regard to when and where the adult individuals of the "bad species" were obtained in nature. Simply, some species can be bred in captivity and some cannot.

As the immediate cause of the death of the animals is usually some kind of parasitic disease, another possible explanation comes into consideration. Some species could have a good immune system, so that they readily come to terms with the new pathogens that they necessarily encounter in captivity, while other species have a **bad immune system**, as a result of which they readily succumb to new pathogens.

However, how is it possible that both groups of species form more or less the same size populations in nature? At first glance it might seem that resistant species should prosper better, not only in captivity, but also in nature, than species with inadequate immune systems. I think that we successfully solved this problem (still over only our second beer). In actual fact, the populations of the successful and the related unsuccessful species need not differ much in their numbers. This is caused to a great degree by parasites which act in nature, amongst other things, as a sort of stabilizing factor, and even out the chances of the successful and unsuccessful. The effectiveness of transfer of parasites from host to host increases with increasing population density. This increase can be very steep. For parasites transferred by direct contact between individuals, the rate of spreading of parasites can be proportional to the square of the population density (i.e. the number of individuals living in a given area), as the number of random contacts between individuals is directly proportional to the square of the population density. However, it follows from this that larger and especially more dense populations of successful species are exposed to a greater parasite burden in nature than the smaller and thinner populations of unsuccessful species. Thus, it can happen that two species that have drastically different viabilities and thus rates of reproduction may differ very little in the sizes and densities of their natural populations. Even this small difference in density is sufficient for some kind of parasite or some kinds of parasites to even out the advantage that follows for the more successful species from its greater viability. The resultant rate of increase in the sizes of the population of both species is finally the same, i.e. zero, see turbidostatic and chemostatic species (Chapter 15). However, species with greater viability achieve this zero rate, i.e. identical rates of reproduction and dying, at a somewhat greater population density.

Why a duck-billed platypus cannot be kept for its wool

Satisfied with the solution to that part of the problem, we ordered another beer and enthusiastically set about solving another, related problem. Could the existence of good and poor quality species explain the fact that only a **minimum number of species have been domesticated** to the present day?

Box 17.1 Domesticated species

As opportunity arises, man keeps a large number of species of animals and grows a large number of species of plants for his enjoyment and for their usefulness. He has managed to domesticate – adapt to the conditions of life in captivity – only a much smaller percentage of these species (and a negligible percentage of the species existing on the Earth). The breeder must carefully adjust the conditions for ordinary species kept in captivity and must try to prepare conditions in the terrarium or greenhouse that are as close as possible to the conditions in the natural environment. In contrast, domesticated species do not require anything of this kind, and their populations are capable of successfully surviving under quite artificial conditions enforced by man. It is another matter whether the domesticated animals are as happy under these conditions as they would be in nature; at the very least, my cat Ferda mostly seems to be.[1]

A well-known and very obvious phenomenon is involved here. There are thousands of bird and mammal species in nature. However, only a negligible percentage of them have been domesticated in the past. This is also true of cultivated plants. Were our predecessors lacking in fantasy, patience, motivation or was this really impossible for some objective reason? What if it is truly possible to domesticate, similarly to successfully keep in terrariums, only a certain small percentage of "good-quality" species and we will not manage with the others whatever we do?

I will not tell you how many beers we drank before we returned to the possible reasons for the differences in the qualities of these species. The reader could come to the quite erroneous conclusion that teachers at the Faculty of Science of Charles University are fond of alcohol. Amongst other things, it is interesting in connection with this phenomenon (we are again talking of good and bad species, not alcoholism amongst the teachers at the Prague Faculty of Science) that there seems to be a quite sharp boundary between good- and poor-quality species. In biology, very few things are completely black or white, mostly various continuous shades of grey are encountered. Consequently, biologists are well trained in recognizing any discontinuities. In the case of good- and poor-quality species, it seems that there is a discontinuity, i.e. that there is quite a sharp boundary. Some species are resistant and survive well in captivity and especially these species are sold and purchased at terrarium markets; on the other hand, some species are almost impossible to keep in captivity and are purchased only by Daniel and other incorrigible adventurers.[2] And there is nothing, or, to be more exact, very little, between the extremes. This discontinuity in the quality of species could be the key to understanding the nature of this phenomenon. Overall, there are few reasons why species should be divided into two separate groups on the basis of the quality of their immune systems. As the immune system consists in an extremely complicated network of mutually cooperating, but frequently quite independently working mechanisms, we should see in nature a more or less connected continuum extending from very poor-quality species to species with very good immune systems. However, such a system is not observed between domesticated and undomesticated species.

And we have it! – frozen plasticity

Does not the reason for the ability of some species to be kept and domesticated lie somewhere else entirely? Could not the theory of frozen plasticity offer an explanation? According to it, there should truly be two groups of species – on the one hand, the smaller group of evolutionarily plastic species and, on the other hand, the large group of evolutionarily frozen species. It is quite possible that evolutionarily plastic species will be more capable of surviving for long periods of time under the extreme conditions of captivity and will be more readily domesticated. Thus, the strange and exotic red junglefowl (*Gallus gallus*), the wild predecessor of the domestic chicken (*Gallus gallus domesticus*) could be a relatively young species that did not have time to freeze evolutionarily, while grouse, partridge and quails could be evolutionarily frozen and thus not useful for the purposes of domestication. This could be tested – for example, by perusal of an atlas of mammals or birds and seeing if domesticated species of animals are suspiciously young and are, for example, part of a group of several close relatives, possibly still crossable species. In the case of plants used in agriculture, these should be primarily species capable of asexual reproduction (e.g. graftable or with tubers of rhizomes) or at least self-pollinating species. I do not know how things are with domesticated animals (however, I have a feeling that sheep and goats are quite willing to cross together and with other wild species and I think I have heard something similar about domestic Bovidae). On the other hand, I think the situation is quite clear in relation to cultivated plants. For example, it is known that the vast majority of varieties of fruit trees were obtained as somatic mutants (i.e. different fruit began to appear on one of the branches of a tree) and they were reproduced further only asexually, e.g. by grafting. Our improvement efforts are much less successful in plants where asexual reproduction is not successful (walnuts, edible chestnuts).

And now it just occurred to me – one of the reasons why biologists have been satisfied to date with the concept of continuous evolutionary plasticity of all the species could be that they performed their research on the very few species of actually evolutionarily plastic species, which they successfully kept in their animal facilities or grew on their test fields. I have not heard that someone would perform a successful selection experiment on a hazel grouse or on an aardvark. It is certainly easy to demonstrate evolutionary plasticity if the experiment is carried out on only those species that have already demonstrated evolutionary plasticity by surviving under conditions in captivity or were even successfully domesticated.

And how did my debate with Daniel on good- and poor-quality species turn out? Great, of course. As the number of marks on our bill increased, we smoothly went on to related topics and with great enthusiasm and the assistance of our colleagues we finally got around to our favorite discussion program "A little while with rumours".

Summary and incitement

The difference between species that can be easily kept in captivity or domesticated and species that cannot be subjected to such restrictions could lie in the fact that the former are evolutionarily plastic and the latter evolutionarily frozen. This should be manifested in the fact that the domesticated species are young in terms of evolution and should be part of a group of mutually crossable species and, for plants, should be species that can be reproduced asexually or at least by self-pollination. In experiments performed in laboratories, the organisms could behave with evolutionary plasticity because the species kept in laboratories are preferentially those species that can stand such treatment, i.e. evolutionarily plastic species. The following chapter will be concerned with the effect of the special genetic architecture of evolutionarily frozen species on the progress of some evolutionary processes, primarily the development of altruistic behaviour.

Footnotes

1. The domestic cat is probably not a typical example of a domesticated animal. I have long had a serious suspicion that cats moved in with people on their own initiative and, if anyone was domesticated in this case, then probably cats domesticated humans. In this case, however, I can report that at least some of the domesticated species are quite satisfied with the result of the whole process.

2. The entire matter is somewhat more complicated. Normal terrarium keepers tend to purchase primarily nice-looking species at markets. Herpetologists (including Frynta) purchase ugly (and pretty) species and consequently have an opportunity to (unfortunately) distinguish between species that can and cannot be kept in captivity. That is if they are observant enough.

CHAPTER 18 Could the theory of evolutionary plasticity explain the formation and maintenance of altruistic behaviour?

The theory of frozen plasticity indicates that the evolutionary response of the population to selection pressure is fast amongst evolutionarily frozen species, i.e. most species that we encounter in nature, but that it is simultaneously short in duration, so that the result is small in its extent. As we showed in the previous chapter, the species that we employ in our experiments can respond much more readily to selection pressures, both because primarily domesticated species are employed in laboratory experiments, and also because our experiments were based on a very small population and thus contained a minimum of genetic variability. This means that our experimental organisms were either originally (i.e. prior to domestication) evolutionarily plastic, or that we brought them into a plastic state by "forcing them through a bottleneck", through the stage of a very small population.[1]

Because of the existence of plastic and elastic stages in the lives of species and because of the difference in the duration of the stages, it can be expected that primarily evolutionarily frozen species should be encountered in nature, with only a smaller number of evolutionarily plastic species. On the other hand, primarily evolutionarily plastic species should be present in our laboratories, as was mentioned above. Plastic and frozen species should respond to selection pressures differently. Frozen species should initially respond faster to short-term pressure (as their gene pools contain sufficient genetic variability from which they can select suitable alleles); however, after the frequency of the individual alleles is deflected from the original equilibrium, the population stops responding to the selection pressure and, after the end of selection, the proportional representation of the alleles and thus the phenotype of the organisms usually returns to the original state. In contrast, evolutionarily plastic species respond more slowly to selection pressure, because they lack genetic variability from which selection could choose and they must rely to a considerable degree on new mutations; however, the selection remains effective over a much longer period and, after it ends or is interrupted, the phenotype of the organism does not return to the original state.

In what other way do plastic and frozen species differ?

Differences in the speed and duration of the evolutionary response to selection pressures are, however, not the only differences in the evolutionary behaviour of plastic and frozen species.

Differences in the **genetic architecture** of evolutionarily plastic and evolutionarily frozen species can be manifested in other ways in evolution. To explain, by genetic architecture I mean not only the amount of genetic variability, but also the way in which the individual phenotype traits are coded, i.e. the genetic nature of the difference between individuals in the population. Amongst evolutionarily plastic species, it can be expected that most of the genetically determined differences between the individuals in the population will be coded by a single gene and that the presence or absence of a single specific allele will be responsible for this. In frozen species, on the other hand, most of the genetically determined differences will be coded by a greater number of genes and the effect of genetic interactions will be important to a great degree and very frequently.

It is not pure chance or a manifestation of my contrariness or perverted liking for accumulation of professional terms that made me use the seemingly superfluous expression "most of the genetically determined differences" rather than "most differences" in both the previous statements. In evolutionarily plastic species, it is almost certain that most of the observed differences between individuals in the population will be of a nongenetic nature, i.e. will be determined by the effect of the environment and will not be inherited from the parents by their progeny (otherwise the individual alleles would have become fixed long ago, or would have been removed from the population by selection). In contrast, amongst frozen species, a substantial part of the differences will be determined genetically and will reflect genetic differences amongst the members of the population. Thus, the differences will frequently be inherited from the parents by their progeny; however, because each of them is usually determined by a greater number of genes and genetic interaction comes into play in their formation, heredity will generally fade out, i.e. they will be transferred from one generation to the next to an ever lessening degree as the genes that originally caused their formation (correctly alleles) are gradually diluted – see Chapter 9.

Might there really be group and inter-species selection?

And thus we arrive at another interesting result following from the theory of frozen plasticity. Amongst frozen species, a very important role can be played by the frequently derided and doubted processes of **group selection** and **inter-species selection**, i.e. processes in which, in the first case, whole groups (herds or flocks) of organisms compete or, in the second case, entire species, and that could theoretically lead to the development of traits advantageous for the group or species and simultaneously disadvantageous for their individual carriers. This could be a substantially greater role than attributed by the contemporary Neodarwinist theory of evolution and than evolutionary biologists are currently willing to admit.

Let me explain. As we mentioned in Chapter 16, the main and most common objection of evolutionary biologists against the role of group and inter-species selection in evolutionary processes consists in the fact that a trait that provides an advantage to a group and simultaneously places the individual that is its carrier at a disadvantage has no chance of spreading and enduring for a long time in nature. Groups (species) in which the altruistic trait spreads would prosper better than groups in which this trait is lacking and the average biological fitness of the members of this group would be greater; however, selfish individuals who do not exhibit this trait and do not behave altruistically, but only enjoy the advantages provided by the presence of altruists, would have the greatest biological fitness within these groups. Selfish individuals would thus have the greatest rates of reproduction and over time would logically force altruists out of the population, to the detriment of the group or species as a whole. The classical theory of evolution admits the possibility of group (and inter-species) selection only where this is a matter of the traits that bring an advantage to the population (species) and simultaneously do not place their carriers at a disadvantage, or where the population has a rapid turnover, and populations frequently disappear and appear again independently from a small number of founding individuals. For the latter case, evolutionary biologists began to use the term "**interdemic selection**", probably so that they would not have to rewrite the textbooks and began to pretend that this is an entirely different phenomenon than group selection. Well, do not believe them; this is group selection whatever they say. However, it entails the problem that it can play a role only at certain rates of dying out and founding of populations and dying and multiplication of their members. Altruistic traits, which are quite frequently encountered in nature, then generally began to be considered to be a product of **kin selection**, assisting genetically related individuals to increase the inclusive fitness of the "altruistic" individual – which is, of course, a form of individual selection, or a manifestation of **reciprocal altruism** – today I will help you (with gritted teeth) so that you will help me tomorrow. Once again, individual selection comes into play rather than group selection.

What does the theory of frozen plasticity have to say about this? I would say that something very important. Here I am not thinking so much about the fact that, in an evolutionarily frozen species, individual selection cannot so easily eliminate an altruistic trait from the population simply because any type of selection is rather ineffective here. The formation of the altruistic trait could be connected with the period of evolutionary plasticity of the species and, although this is a very short period of time compared to the duration of the subsequent period of evolutionary freezing (remember that this corresponds to about 1–2% of the time of existence of the species), it is still more than sufficient for forcing out the carriers of the altruistic traits as a result of individual selection. I have in mind something far less conspicuous and simultaneously far more important.

I will not test the patience of the reader any further. The main reason why it is far from necessary to abandon the concepts of group or inter-species selection lies in the fact that an individual trait, for example altruistic behaviour, is usually determined by a greater number of genes and consequently the inheritability of most traits fades out. If, for example, altruistic behaviour is determined by the presence of four genes at four different sites in the genome, where two of the participating genes can replace one another in their effects, then there is not a very great difference between the probabilities that an altruist will be born in the family of an altruist or of a selfish person. Altruists emerge from the population as if by chance in families that are completely unrelated and have different phenotypes, i.e. individuals with quite different behaviour, with a probability that is determined only by the proportion of the alleles in the entire population. **Thus populations (species) can compete for the greatest average biological fitness of their members and those that have the greatest proportion of the relevant alleles, as a result of which the greatest number of altruists will be formed (emerge by chance), will win in this competition. Thus, group and inter-species selection can occur in nature in favour of altruistic traits (because the percentage proportion of alleles in the population is inherited from one generation to the next) and its results cannot be cancelled out by individual selection because the trait itself, altruistic behaviour, is not inherited.**

Farewell to Eugenics!

Thus, Mother Theresa could have been born in the family of Usama bin Ladin or vice versa.[2] The fact that children later copy some patterns of behaviour of their parents or protest against them (or do both at once) is true, but is rather a special feature of our species and is not connected with the transfer of genes. The consistent transportation of criminals from England to Australia did not have any substantial impact on the level of criminality in that continent, or in England. Even if some part of criminal behaviour could be determined genetically, the roulette of sexual reproduction meant that, probably already in the second generation, the proportion of altruists and selfish persons in Australia and in England returned to the original levels.

The individual genes that, in a certain combination, determine the formation of a trait that is advantageous from the standpoint of the population and disadvantageous from the standpoint of the individual can also determine the formation of other traits, which are different for each gene. And it is selection for the occurrence of these other traits that can maintain them permanently in the population, i.e. can regularly remove mutants in which the relevant genes are damaged as a consequence of mutation. This kind of selection (called **negative or purifying selection**) can, of course, also be active in evolutionarily frozen species. Once again, I can offer only a hypothetical example. Imagine that the tendency of Jackdaws to jointly attack predators in the vicinity of their nests (please take note that I do not read only studies on molecular biology but also the books of Konrad Lorenz) is determined by the simultaneous presence of two alleles on two different genes. Each of these alleles simultaneously determines one more trait, for example one is responsible for care for the young and the second for curiosity (an ethologist would describe this as exploration activity). If, in the given environment, it is advantageous for the individual to care for its offspring and to exhibit curiosity, then the number of altruists present in the population (which will participate in defence against predators instead of selfishly waiting to see how things turn out, and those who the predator captures) will increase, or at least will not decrease.

Maintenance of the genetic predisposition for a complicated trait (e.g. for altruistic behaviour) in the population in such a disassembled state by selection in favour of completely different traits can have an unpleasant consequence for the work of phylogeneticists. Similarly as a certain trait

Box 18.1 Determination of phylogenesis

The most important method of determining phylogenesis is based on gradual connecting of species that share new evolutionary features, called **apomorphic traits** (the opposite of an apomorphic trait is a **plesiomorphic trait** – the original evolutionary form of the trait). Traits that are so complicated that it can be assumed that they were formed in evolution only once and that species that share them did not form the trait independently, but inherited them from a joint predecessor, can be considered to be useful apomorphic traits for phylogenetics. If two species A and B share ten apomorphic traits, but share only seven apomorphic traits with a third species C, it can be assumed that species A and B branched off from a joint predecessor in evolution later than that predecessor from the species that was also a predecessor of species C.

may emerge in unrelated individuals, it can most probably emerge after a longer period of time in mutually unrelated species whose members and immediate predecessors did not exhibit this trait. However, this can substantially distort the results of phylogenetic analysis, as it could lead to the wrong conclusion that species that need not have any connection are mutually related.

I quite understand that the above discussion and actually the entire theory of frozen plasticity is bad news for the proponents of eugenics.

Not only is any form of selection in the case of an evolutionarily frozen species hopelessly ineffective but, basically, there are not even any criteria according to which this selection can be made.[3] Some individuals behave altruistically and some selfishly, some are extremely intelligent and others extremely stupid, some are beautiful and strong and others are ugly and sickly. However, there is no evidence suggesting that, if we prevent one stupid individual from reproducing, we remove more genes for stupidity than if we prevent any other randomly selected individuals from reproducing. Of course, in the case of traits determined by the presence of one gene, selection can be effective; however, there are probably very few of these traits and a great proportion of the relevant alleles were most probably formed in the population quite recently as a consequence of new mutations. Thus, it is probably exaggerated to fear that the genetic

Box 18.2 Eugenics

Eugenics attempts to improve mankind through the methods of genetics or rather breeding. Negative eugenics attempts to eliminate from the population or exclude from the process of reproduction the carriers of alleles determining the formation of undesirable traits, while positive eugenics attempts to promote the reproduction of desirable traits. In the past, eugenics could use only methods that were more or less unacceptable from an ethical standpoint, specifically killing, interning or sterilizing the carriers of undesirable traits. At the present time, prenatal screening or selection of germ cells, zygotes or early embryos could be employed. Thus, programs concerned, e.g., with eliminating alleles coding some serious genetic defects, become ethically more acceptable. On the other hand, at least one fundamental risk remains here. So far, we are not capable of answering the question of what we cause when, through removing the relevant alleles from the population, we prevent, e.g., the birth of some forms of mentally or physically damaged individuals. What if these alleles in a different genetic context (i.e. in combination with other alleles) simultaneously cause the birth of geniuses or resistance to tuberculosis?

composition of future populations would deteriorate substantially because the methods of modern medicine permit individuals bearing detrimental alleles to survive (and reproduce). On the other hand, systematic searching for the carriers of the relevant recessive alleles and subsequent use of the methods of prenatal diagnosis and assisted reproduction can be quite effective in the case of some hereditary diseases.

Now I have probably irritated a number of people who do not agree with prenatal diagnosis and assisted reproduction. This was probably not very sensible from the standpoint of the sales potential of the book and thus, implicitly, from the standpoint of my biological fitness. On the other hand, the end of the book is near and thus you will probably not try to return it at this point, if only because you spilled juice on it, or what is that stuff on page 67. Moreover, in modern times, heretics are no longer burned at the stake and prophets are not stoned to death. Returning to the subject of the chapter, even if they were burned at the stake and stoned to death, mankind can remain easy. As follows from the previous paragraphs, for each burned or stoned author of a controversial book, another one will break off somewhere else.

Summary and incitement

To summarize. The genetic architecture of evolutionarily plastic and evolutionary frozen species probably does differ. In plastic species, there will be a greater percentage of traits coded by a single gene, while in frozen species a greater percentage of traits will be coded by a large number of genes and genetic interactions will play a greater role here. Amongst other things, this means that altruistic traits occurring as a consequence of group or inter-species selection can survive much more easily in frozen species. Here, altruism is not inherited from one's parents; altruists emerge (break off) by chance in mutually unrelated individuals with a probability dependent on the proportion of the particular allele in the population. The numerical proportion of alleles that lead to the formation of an altruistic trait in only a certain combination is inherited in the population from one generation to the next, thanks to which group selection can occur. A single gene can affect a number of various traits. The individual alleles, whose specific combination causes, e.g., the formation of altruistic behaviour, can thus be maintained in the population independently by selection for the presence of quite different traits. As a consequence, a particular complicated trait can emerge repeatedly, not only in unrelated individuals that do not carry it themselves but, after a long time, even in unrelated species. This can distort the results of phylogenetic analysis. In the last chapter of the book (do not celebrate prematurely, I might just write another volume), I will attempt to explain the reasons why I wrote this book and what its purpose was intended to be.

Footnotes

1. The bottleneck effect is a common term in population genetics. If, for a short temporary time, the population is reduced to a fraction of its original size, this is subsequently reflected in the genetic structure of the population. The most important consequence of the passage of a population through a bottleneck is the disappearance of most rare alleles from the gene pool of the population. However, if the reduction in the size of the population is actually short-term and is followed by a period of rapid population growth back to the original value, the more common alleles will not have had time to disappear (i.e. the alleles that are maintained in the population by frequency-determined selection or selection in favour of heterozygotes). In addition, in the period of increasing population, the action of selection is limited so that a great many newly formed, slightly detrimental alleles can appear in the population.
2. I used these persons only as widely employed symbols of good and bad properties. I do not intend to get into a discussion of whether Bin Ladin caused more suffering through his terrorism or Mother Theresa through her rejection of contraception.
3. Humans are a relatively young species and thus it cannot be excluded that they are still, to a certain degree, plastic. On the other hand, they are not completely young; their period of existence is definitely ten times longer than the estimated duration of the plastic phase of a species (i.e., say, 10 000-20 000 years).

Chapter 19 A few words in conclusion

Two calls to the nation played an important role in recent Czech history. It must be admitted that the information value of their titles was rather low in both cases. During the Prague Spring in 1968, this was the Two Thousand Words manifesto and, in 1989, the Several Sentences petition. In selecting a title for the concluding chapter of the book, I decided to continue this modern national tradition, manifested by a tendency to designate important texts by names that have practically no information content.

Evolutionary biology has gone an enormously long way over the past 150 years. From tolerant Darwin's evolutionary pluralism (selection, and probably a great many other processes, are active in evolution), through the somewhat reticent attitude of the professional public towards natural selection at the beginning of the 20th century, through the period of Neodarwinism, in which natural selection was considered to be fundamental and basically the only important source of evolutionary change, to the present time of a sort of hidden renaissance of evolutionary pluralism where, under the cover of official teaching, modern evolutionary biology is developing literally in all possible directions. The modern (one could almost say post-modern) tolerant approach, however, entails one great disadvantage. Because of this tolerant approach, it is very difficult to orient oneself in the field from the outside, it is very difficult to recognize promising new ideas and it is difficult to decide on promising directions for research. It is rather a mechanical concept, but if development is occurring in all directions, then the field itself, at the very least its centre, is necessarily not going anywhere. In a great many fields of scientific activity, such a state does not matter and could even be ideal from the standpoint of scientific progress and freedom of research. However, it entails a great danger in the area of evolutionary biology. A special feature of evolutionary biology and, as some people are fond of saying, its greatest puzzle is the fact that almost everyone thinks they understand it. I would like to add that its other puzzle is that a great many people, in fact including a number of biologists, are wrong in this. The Darwinist principle of natural selection is truly an elegant mechanism that can be employed to explain natural phenomena that are otherwise difficult to understand. If you understand the basic principle of natural selection (and almost everyone is capable of this, although frequently in rather distorted form), then you can very easily erroneously come to the conclusion that you understand all of evolutionary biology. And thus it frequently happens that biologists who are professionals in their original field (in the better case, I had better not elaborate the worst case) begin to devote themselves to evolutionary biology, although they are really only self-taught in the field of evolutionary biology. (In order to deprive the malicious reader of a weapon, I would like to point out that I, too, did not study evolutionary biology, but cellular biology and physiology.) Consequently, it very frequently happens that, not only is that which is already known discovered again (this would not be so terrible, and happens in a great many areas of human effort on a daily basis), but also what has already been discovered is forgotten.

The theory of interallelic selection (the selfish gene) and the theory of evolutionarily stable strategies have been known in evolutionary biology since the 1970s. The model of punctuated equilibria has been known for approximately the same time. The fact that these three things are interconnected and, together, indicate that the contemporary, still Neodarwinist understanding of evolutionary biology is most probably erroneous from its very foundations and requires fundamental re-examination, has not penetrated into the consciousness of evolutionary biologists (and even less into textbooks on evolutionary biology).

It is true that a few geneticists have been studying the basic assumptions of the theory of frozen plasticity for the last 30 years, probably out of inertia, by studying the response of genetically variously variable populations to selection pressures (I would not want to act on their consciences but, from outside, it frequently seems like they have long forgotten why they are actually doing it), but their works, which are very frequently published in very good genetics journals[1], generally do not draw much attention from other biologists. The fact that the results of their experiments on house flies, fruit flies or flour beetles of the *Tribolium* genus (these are small black or reddish beetles that get into the flour and, if you accidentally bite into one, taste like tar) could shake the very foundations of contemporary evolutionary biology is something that no one realizes. The four main spiritual fathers of the theory of frozen plasticity, Ernst Mayr, William D. Hamilton, Stephen J. Gould and John Maynard Smith all died at the beginning of the 21st century (I certainly cannot reproach 101-year old Mayr for leaving the battle field prematurely, but the other three really annoyed me by their irresponsible behaviour. And I am not even mentioning that they left an unfortunate vacuum in biology.) And now we are faced by the quite serious danger that a Copernicus-type revolution in evolutionary biology will be put off indefinitely for lack of general interest. If the theory of frozen plasticity is correct, and I would bet that it is, then it will be rediscovered sooner or later. However, it cannot be guaranteed that it will not be alternately rediscovered and reforgotten. The results of experiments with flour beetles do not look very "sexy" and, even if they confirmed the theory of frozen plasticity a hundred times over (which has not happened yet – see Chapter 13), the authors of textbooks in evolutionary biology will probably not learn of this, or will at least act as if they do not know about it.

The book that you are finishing is an attempt to combat this fate and push the theory of frozen plasticity into evolutionary biology by the side door. I have already tried coming in by the main door, i.e., by publishing a normal scientific article in a professional journal. The article was finally published in the not-very-well-known journal Rivista di Biologia in 1998 and, as could be expected, received no response at all. (In fact, because of my enormous conceit, I consider this to be independent confirmation of the validity of the presented theory. I have found that the number of references to my articles is inversely correlated to the importance of the published results. In this respect, the article on the subject "Does the cell perform isoelectric focusing?"

which I published in 1990 and which has also never been cited, is apparently very promising.[2]) I do not expect the book to convince evolutionary biologists to abandon their favorite theories in favour of the theory of frozen plasticity. They will probably tend to consider it to be a sort of unimportant intellectual puzzle that need not be taken seriously. However, even if they found that they could work their way through it in this form and read right to the very end, I would be glad. For example, it could instigate them to occasionally pose a question that they would otherwise not pose.

I have no idea how biologists in other fields will view the book. I hope that, at the very least, they will take away the impression that evolutionary biology is certainly not a closed chapter and that a number of quite fundamental questions remain to be resolved. And I hope that at least some of them, after reading the book, or even during reading it, will strike their foreheads and say: "Good Lord, if that were true, then that would explain why ..., that ... and how .... And that could actually be tested."[3]

The book is intended primarily for young people, secondary school and university students, who are only now looking for the area of their future profession. Some of them could end up as my future colleagues and, sooner or later, replace the contemporary generation of evolutionary biologists. Instructed in Kuhn's theory of the development of science, I am well aware that the only known effective way of replacing one long-term successful theory by a new one is to let the proponents of the old theory die a natural death and let a generation of proponents of the new theory grow up in their place. And it is necessary to work on this sufficiently far in advance. I would be very happy if my book were to contribute to the emergence of a generation of biologists for whom it will be quite natural to question the basic paradigms of Neodarwinism, i.e. ask whether all organisms (i.e. both sexually and asexually reproducing species) can change through the action of natural selection throughout their existence. I cannot predict how they will answer this question, but I am optimistic in this respect.

I hope that I have not simultaneously discouraged young readers from following a scientific career in that I occasionally intentionally revealed some of the externally hidden secrets about the functioning of modern science (here I would like to cite one of my colleagues, who was so kind as to read and comment on my manuscript: "Occasionally? All the time! It was an obsession with you!"). I tried to do this in the lightest way possible (I quote: "Unsuccessfully!"). What I intended was to show that, although modern science does not function completely without problems, it is hard to think of another, similarly efficient system of accumulating knowledge. A number of the problems in the manner, management and self-management of contemporary science can probably not be eliminated and we must learn to live with them. There is no point in running down something that we cannot replace with something else; however, it can be useful and healthy (at least for us, ourselves) to learn to make fun of it. It is quite possible to love science and simultaneously not to take science, scientists and, first of all, oneself very seriously. As my favourite author, Jan Werich once said "One shouldn't make a science out of anything. Not even of science."

Footnotes

1. I would like to apologize to the non-specialist reader, but a rather dry (and very incomplete) list of genetic works related to the theory of frozen plasticity will be given here. In contrast to the rest of the text, here I am intentionally giving full citations so that it is at least partly apparent to what each article is related. Carson, H.L. and Templeton, A.R., Genetic revolutions in relation to speciation phenomena: the founding of new populations. *Annual Reviews of Ecology and Systematics* 15: 97–131, 1984; Templeton, A.R., The theory of speciation via the founder principle. *Genetics* 94: 1101–1138, 1980; Bryant, E.H.S., McCommas, A. and Combs, L.M. Morphometric differentiation among experimental lines of the housefly in relation to a bottleneck. *Genetics* 114: 1213–1223,1986; Goodnight, C.J., On the effect of founder events on epistatic genetic variance. *Evolution* 41: 80–91, 1987; Whitlock, M.C., Phillips, P.C. and Wade, J.M., Gene interaction affects the additive genetic variance in subdivided populations with migration and extinction. *Evolution* 47: 1758–1769, 1993; Bryant, E.H.S., McCommas, A., and Combs, L.M., The effect of an experimental bottleneck upon quantitative genetic variation in the housefly. *Genetics* 114: 1191–1211, 1986; Galiana, A., Moya, A. and Ayala, F. J. Founder-flush speciation in *Drosophila pseudoobscura*: A large-scale experiment. *Evolution*47: 432–444, 1993; Cheverud, J.M., and Routman, E.J., Epistasis as a source of increased additive genetic variance at population bottlenecks. *Evolution* 50: 1042–1051, 1996; Goodnight, C.J., On the effect of founder events on epistatic genetic variance. *Evolution* 41: 80–91, 1987; Katz, A.J., and Young, S.S., Selection for high adult body weight in *Drosophila* populations with different structures. *Genetics* 81: 163–175, 1975; Lopez-Fanjul, C., and Villaverde, A. Inbreeding increases genetic variation for variability in *Drosophila melanogaster*. *Evolution* 43: 1800–1804, 1989; Naciri-Graven,Y. and Goudet, J., The additive genetic variance after bottlenecks is affected by the number of loci involved in epistatic interactions. *Evolution* 57: 706–716, 2003; Day, S.B.and Bryant, E.H., The influence of variable rates of inbreeding on fitness, environmental responsiveness, and evolutionary potential. *Evolution* 57: 1314–1324, 2003.
2. On the basis of some information about cell physiology, I came to the conclusion while I was still at university that eukaryotic cells (whose internal volume is usually 2 000–10 000 times greater than the volume of bacterial cells) cannot depend on diffusion in the transport of molecules, but move proteins and possibly also low-molecular substances from one place to another or concentrate them in certain regions of the cytoplasm using isoelectic focusing. (Stated simply – if a molecule with an electric charge is exposed to a voltage field in an environment in which there is a pH gradient, they move to places with a pH at which their electric charges will equal zero.) In my paper published about 10 years later, I collected evidence for this model and also demonstrated the consequences that this could have for cell physiology and biochemistry and especially I calculated that the cell has sufficient energy for effective focusing. The fact that I made an unfortunate error of several orders of magnitude in my conclusions (against my hypothesis) and was forced to publish a Corrigendum in the same journal seven years later is another matter (sad for me, maybe quite funny for some others). The article can be found in *BioSystems* 24: 127–133, 1990, and the *Corrigendum* in *BioSystems* 37: 253, 1996. (After reading this note, a certain colleague stated the serious suspicion that I discovered the entire theory of frozen plasticity only so that I could subsequently promote my theory of intracellular isoelectric focusing. I hereby state that this is a shameless defamation – in actual fact, I wanted this to be a way of promoting our discovery of the protective effect of the Rh factor against lowered performance in persons infected by latent toxoplasmosis. In any case, decide for yourself. Why else would I bring it up right in the second chapter …)
3. It seems that it really works this way. One of the unofficial reviewers of the book, Stanislav Komárek pointed out the possibility that the theory of frozen plasticity could explain the reduced biological fitness of most "improved" breeds of animals and their spontaneous return to the original wild forms; another of the reviewers, David Storch pointed out that the various ages of frozen species could explain the enormous differences in the numbers of members of the individual species and Petr Baum proposed a new definition of a biological species: (a set of individuals sharing an identical gene pool in the time between two periods of evolutionary plasticity). And see also the Footnote 12 in Chapter 16.

Index of texts in Grey Boxes

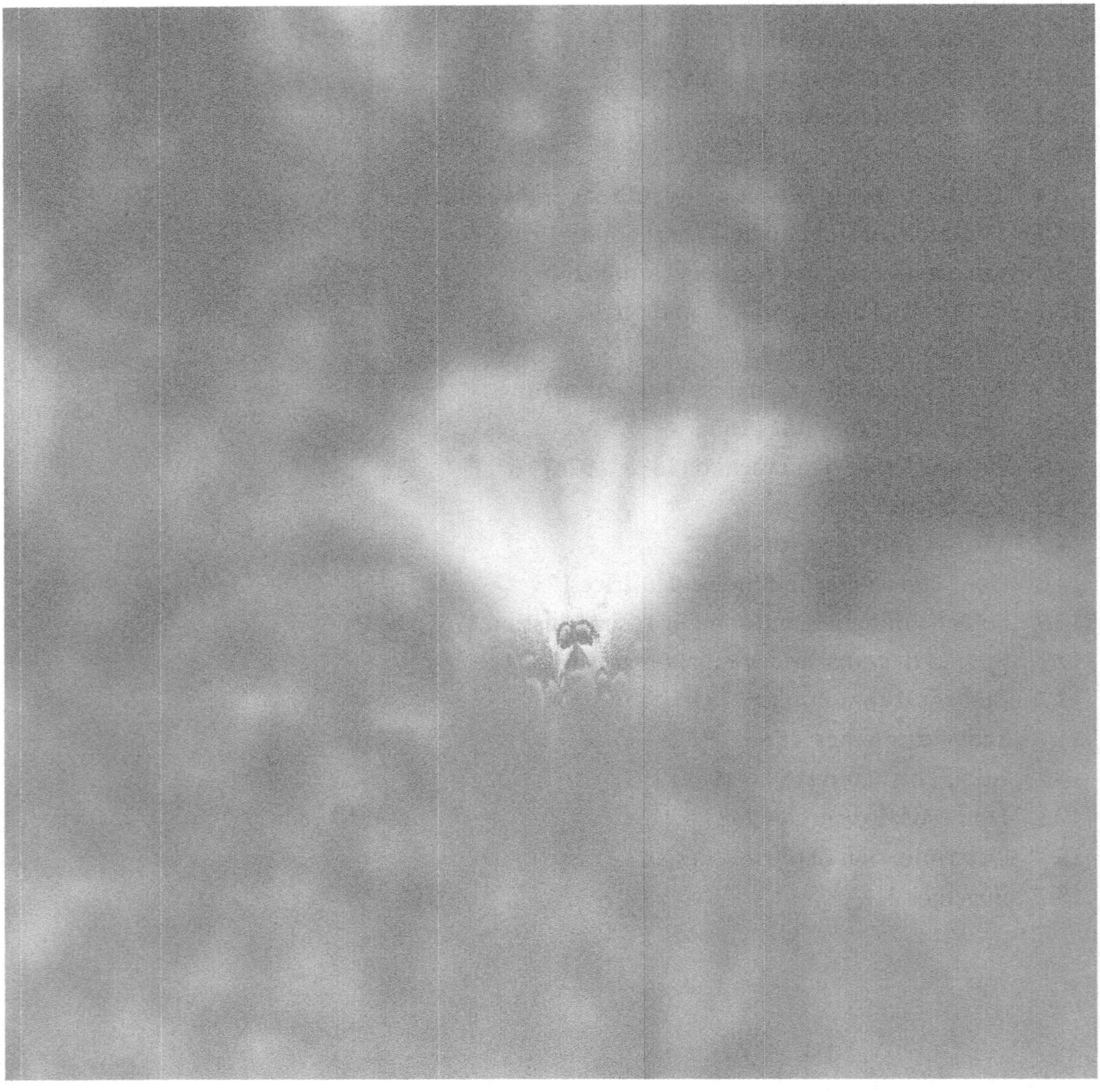

Index

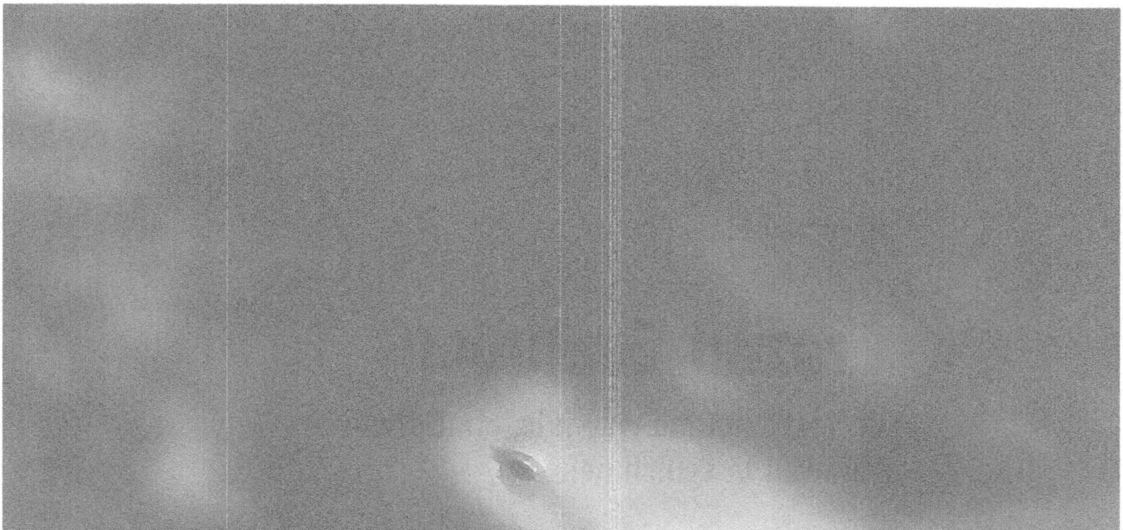

Jaroslav Flegr

Frozen Evolution

Or, that's not the way it is, Mr. Darwin
Farewell to selfish gene

Translated from Czech original: *Zamrzlá evoluce aneb, Je to jinak, pane Darwin.*

Title: Frozen evolution
Subtitle: Or, that's not the way it is, Mr. Darwin – Farewell to selfish gene
Author: © Jaroslav Flegr
Translation to English: Madeleine Štulíková
Editor: Jana Steinerová
Publisher: Charles University in Prague Press
Book Design and Layout: Vladimír Fára
Coverpage illustration: © Pavel Rut
Publisher: Charles University in Prague, Faculty of Sciences
Published: Prague, 2008
Edition: 1st
Printed by: BookSurge, N. Charleston, USA.

ISBN: 978-80-86561-73-8

Made in United States
North Haven, CT
27 August 2026

17261335R00124